Introduction to the Physics of the Cryosphere (Second Edition)

Online at: https://doi.org/10.1088/978-0-7503-3647-5

Introduction to the Physics of the Cryosphere (Second Edition)

Melody Sandells
Geography and Environmental Sciences, Northumbria University, Newcastle upon Tyne, Tyne and Wear, UK

Daniela Flocco
Dipartimento di Scienze della Terra, dell'Ambiente e Risorse (DiSTAR), Università degli Studi di Napoli Federico II, Naples, Italy

IOP Publishing, Bristol, UK

ISBN 978-0-7503-3647-5 (ebook)
ISBN 978-0-7503-3645-1 (print)
ISBN 978-0-7503-3648-2 (myPrint)
ISBN 978-0-7503-3646-8 (mobi)

DOI 10.1088/978-0-7503-3647-5

Version: 20221201

IOP ebooks

British Library Cataloguing-in-Publication Data: A catalogue record for this book is available from the British Library.

Published by IOP Publishing, wholly owned by The Institute of Physics, London

IOP Publishing, No.2 The Distillery, Glassfields, Avon Street, Bristol, BS2 0GR, UK

US Office: IOP Publishing, Inc., 190 North Independence Mall West, Suite 601, Philadelphia, PA 19106, USA

"Nothing in life is to be feared, it is only to be understood. Now is the time to understand more, so that we may fear less." – Marie Curie

We would like to thank our mentors and colleagues who inspire us, taught us how to look for new perspectives and helped us believe in ourselves:

Kathie Bowden, Richard Essery, Daniel Feltham, Ken Golden, Robert Gurney, Ellie Highwood, Margaret Jewell, Danny Marks, Antonio Memoli, Liz Morris, Don Perovich, Nick Rutter, Stefano Sosti and Giancarlo Spezie, Leanne Wake and Adam Winstral.

Contents

Preface

Introduction to the Physics of the Cryosphere is intended for graduates with a numerical sciences background, particularly those who are heading toward post-graduate study or are generally interested in environmental physics. Conservation equations underpin the physics encompassed in this book, although the interesting part comes in how the necessary variables and boundary conditions are defined to be able to simulate changes in the cryosphere. Phase changes between ice, liquid water and water vapour also come into play.

There are many different components of the cryosphere, for example glaciers, ice sheets, sea ice, lake ice, permafrost and snow. Why these are important and how they can be observed are described in chapter 1. The physics that is common to all of these is described in chapter 2. Each component has features and processes that are unique or more relevant, so these are described in individual chapters. Permafrost is comprised of ice within soil and stones, the thawing of which can lead to land surface collapse, as described in chapter 3. The microstructure of snow continually evolves, yet governs energy transfer as described in chapter 4. Land ice, in chapter 5, is subject to large pressures that can cause melt at the base and reshape the underlying topography. Sea ice has a saline content which affects the thermodynamics, and can also be fast moving and so requires a consideration of momentum, as described in chapter 6. In this second edition, chapter 7 extends the physics to the ocean and atmosphere in the high latitudes, and how they link to the mid-latitudes through circulation patterns.

Although we are part of a vibrant research community, there are many more questions that need to be answered. The final chapter gives a very brief overview of the political challenges and areas of research growth. We hope you will join us and increase our understanding of the physics of the cryosphere.

Note: For readers of the print edition of the book, please find links to various resources here: https://github.com/mjsandells/cryobookurls.

Acknowledgement

We would like to thank our colleagues, friends and family who contributed material for this book: Michele Barone, Sarah Boon, Peter Chaerle, Harry Clayton, Sabrina Cortese, Ronald Daanen, Ian Davenport, Dany Dumont, Sinead Farrell, Kenneth Libbrecht, Julien Meloche, Don Perovich, Jennifer Petrzelka, Ixia Sandells, Arvids Silis, Ina Timling and Peter Webley. We are hugely grateful to be surrounded by brilliant and inspiring people who keep our research interesting and relevant.

Author biographies

Melody Sandells

Melody Sandells studied Physics at Imperial College, London before undertaking her PhD at the University of Reading and postdoctoral work at University College London. She is now an Associate Professor at Northumbria University and is one of the developers of the Snow Microwave Radiative Transfer model. Her research interests include earth observation of snow and soil moisture, snow physics, snow vegetation radiative interactions, microwave remote sensing and hydrology.

Daniela Flocco

Daniela Flocco studied Environmental Sciences and a PhD at the Scott Polar Research Institute at the University of Cambridge. She carried out her postdoctoral research on sea ice thermodynamics at the Centre for Polar Observation and Modelling in London and then in Reading. She moved on to the National Centre for Atmospheric Science, working on a project on sea ice predictability. She now works at the University of Naples Federico II on regional microclimates.

IOP Publishing

Introduction to the Physics of the Cryosphere (Second Edition)

Melody Sandells and Daniela Flocco

Chapter 1

Observing the cryosphere

The *cryosphere* is anywhere on Earth with ice in some form (figure 1.1, 1.2). Ice can be a solid mass on land, like the ice sheets of Antarctica and Greenland, or as flowing glaciers, such as the Mer de Glace glacier in the French Alps. It can be found floating as lake or river ice or subject to ocean thermodynamics as sea ice. As snow it can take many stunning crystalline forms and change extremely rapidly. Embedded within the soil as permafrost or other periglacial landforms, freeze-thaw processes can alter and shape the landscape over timescales from seconds to millenia. This chapter describes the cryosphere today and introduces basic physics behind Earth observation techniques used to monitor changes in the cryosphere. It also dives into climate records and describes how we know the cryosphere has changed in the past. Lastly, we look at the possible prognosis for the cryosphere in the future.

1.1 Components of the cryosphere

Formation, growth and reduction of the cryosphere is driven by changes in energy input to the Earth's surface. Within the atmosphere, cooling at cold temperatures will cause water vapour to condense into solid form, leading to the formation of snow crystals. As soon as snow makes contact, it modifies the energy balance of the land and sea surface. *Permafrost* is ground that has been at a temperature of 0 °C or below for two or more years and its thickness depends on the snow cover above it. *Snow* is a complex matrix of air and ice. Due to the trapped air, the snow acts as an insulator, which is beneficial for small mammals and invertebrates that live in it but may inhibit growth of permafrost that would otherwise be exposed to cold air temperatures.

Over time the snowpack changes in structure and densifies. A *seasonal* snowpack will not last the summer, but if the snow survives a season it becomes *firn* and may ultimately densify to form ice. *Glaciers* are flowing bodies of ice covering 0.1 km^2 or more of the land surface. Ice masses that cover a mountain or mountain range, or land surface up to 50 000 km^2 are known as *ice fields* if they are constrained by the

doi:10.1088/978-0-7503-3647-5ch1

Figure 1.1. Nighttime Scandinavian snow as seen from the International Space Station. Image ISS043-E-86375, courtesy of the Earth Science and Remote Sensing Unit, NASA Johnson Space Center.

Table 1.1. Areal extent of cryospheric components. Adapted from Vaughan (2013).

Cryospheric component	Areal extent (10^6 km^2)
Seasonal frozen ground	48.7
Permafrost	13.3–17.7
Seasonal snow	1.9–45.2
Glaciers	0.7
Greenland ice sheet	1.8
Antarctic ice sheet	12.3
N. Hemisphere lake and river ice	1.6
Antarctic ice shelves	1.6
Antarctic sea ice	2.9–18.9
Arctic sea ice	6.2–14.1

topography, and *ice caps* if they are not. A particular feature of ice caps is a dome-like shape, where the ice flows away from the highest point. *Ice sheets* are bodies of ice covering more than 50 000 km^2 of the land surface; *ice shelves* are attached to land, as they are formed as a result of land ice motion flowing into the sea, but are floating on the water. *Icebergs* are sections of freshwater ice that have broken away from the land ice. *Sea ice* is formed from frozen sea, and therefore differs from land ice because it contains salt. The areal extent of cryospheric components are given in table 1.1. Seasonal frozen ground has the largest spatial extent, with a similar portion of the land surface covered by seasonal snow at its maximum. Snow and sea ice are the most dynamic, with large seasonal variability (Foster *et al* 2005), and thus are more sensitive to climate change. Permafrost, ice sheets and shelves are much more stable as they are sensitive to changes over longer periods of time. Glaciers form the smallest spatial extent, which is decreasing with climate change. An overview of the location of the elements of the cryosphere is shown in figure 1.2.

Figure 1.2 shows the global distribution of the relatively static components of the cryosphere: it exists at all latitudes for some portion of the year. Experience different parts of the cryosphere with this NASA video. To explore the cryosphere in more detail, visit the National Snow and Ice Data Center, where there are many educational resources but who also host data used in cryospheric research.

Annual variability in snow and sea ice area and concentration is illustrated in figure 1.3 for the Northern and Southern Hemispheres, and compares coverage in September 2021 with the March distribution in 2022. These images were formed from the NASA Near-Real-Time SSM/I-SSMIS EASE-Grid Daily Global Ice Concentration and Snow Extent dataset (Brodzik and Stewart 2016), derived from the measurement of naturally emitted radiation at microwave frequencies. In March, sea ice and snow extents are largest in the Arctic but small in the Antarctic. In September, the extents are reversed as the Southern Hemisphere experiences its winter maximum. A similar dataset reveals how Arctic Sea Ice has changed since 1978. Data from the GlobSnow project have been used to show snow mass changes since 1992.

Changes in the weather and longer-term changes in the climate result in evolution of the ice and/or melt. In turn, these changes affect ecology, the economy and human activities. There are a multitude of reasons why it is important to observe the state of the cryosphere, monitor changes and to understand its behaviour.

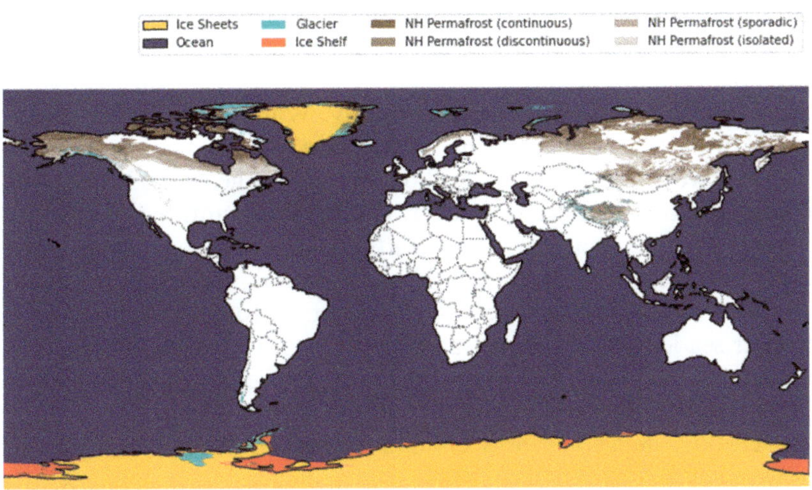

Figure 1.2. Global distribution of ice sheets, ice shelves, permafrost (Brown *et al* 2002) and glaciers (GLIMS Consortium 2005, Raup *et al* 2007).

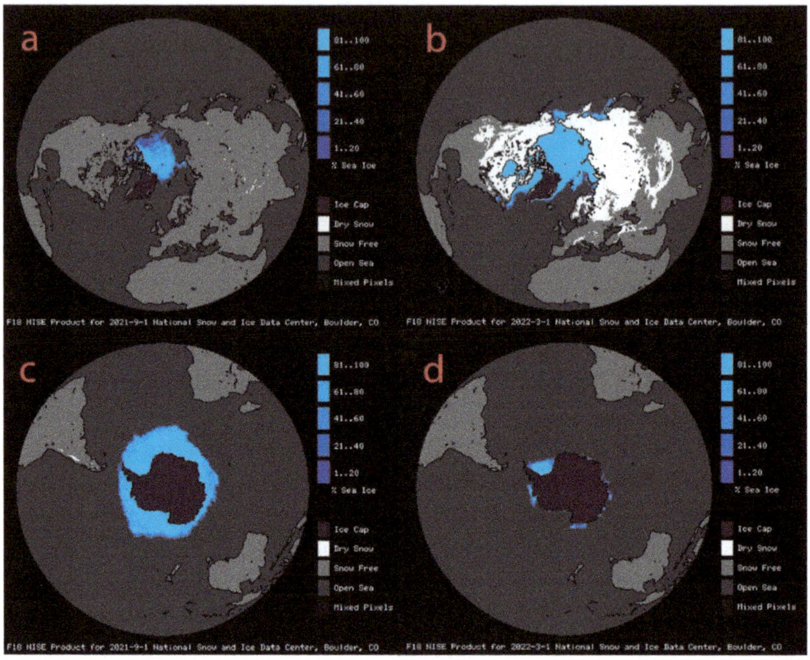

Figure 1.3. Seasonal changes in sea ice and snow cover for the Northern Hemisphere (top) and Southern Hemisphere (bottom). Images are shown for the 1 September 2021 (a) and (c) and for the 1 March 2022 (b) and (d). Data are from the National Snow and Ice Data Center (Brodzik and Stewart 2016).

1.2 Importance of cryospheric knowledge

Cryospheric knowledge has many different impacts: social, economic and ecologic, although a monetary value of the cryosphere is not easily quantified. From a societal point of view, ice and snow are an important source of water; Barnett *et al* (2005) estimated that more than a sixth of the global population needed melt from snow and glaciers for their water supply. Efficient management of this resource requires knowledge of the timing and volume of supply. Water from melt can be used for many things: drinking, manufacturing, agriculture, hydropower and thermoelectric power. Snow is worth more than a trillion dollars to the United States as a water resource, provider of hydroelectric power, its use in agriculture but also accounting for risks due to climate change (Sturm *et al* 2017).

Some effort has been put into quantifying the economic impact of the changing cryosphere. The changes are not necessarily negative from a financial point of view: with a smaller sea ice extent, more transport options become available. On Friday, 27 September 2013, a freighter navigated the Northwest Passage for the first time, ironically on the day the first section of the IPCC 5th Assessment Report (IPCC 2013) was released, describing how climate change effects are more certainly than ever attributed to the actions of mankind. Thus climate change offers an apparent economic benefit for the shipping industry. Conversely, warmer conditions also lead to permafrost thaw, which damages ice roads and other infrastructure, affecting the stability of exploration platforms. The costs of maintaining infrastructure as permafrost thaws is estimated to be in the region of tens to hundreds of trillions of U.S. dollars (Hope and Schaefer 2016); see chapter 3. Other impacts from a changing cryosphere include coastal erosion, sea level rise, tourism and insurance risks (Key *et al* 2007).

Thawing permafrost: human health hazard
Kivalina, Alaska, is a small village in a vulnerable coastal position. Loss of sea ice, which has traditionally protected the village against coastal erosion, and an increase in storms means that it may be uninhabitable by 2025 (López-Carr and Marter-Kenyon 2015). It is estimated that the cost of relocating Kivalina would be US$95–125 million (Fang *et al* 2018), although the value reported in the media was around four times higher.

Cold regions are hazardous, and a good understanding of the physical behaviour of the environment can help monitor and mitigate the risks. For example, snow avalanche risk is assessed by an examination of the meteorological data and digging of snow pits. Evidence of poorly bonded layers indicate that it might not be wise to ski. One of the most memorable conference presentations comprises a video from the Swiss Institute for Snow and Avalanche Research Vallée de la Sionne avalanche test site, which has an observation bunker on one side of a valley and a video system and radar to capture avalanche events on the other side in order to understand mass flow mechanisms (Dufour *et al* 2000). The video starts from the initiation of an avalanche with explosive charges through to the full flow of the avalanche down the valley. It becomes clear from the language and tone of the commentators the precise moment at which the researchers realise that the avalanche is not going to stop at the bottom of the valley. It took four hours to dig the researchers out of the bunker, which was under five metres of snow. As the snow was very dense, the rescuers broke all of their metal shovels and ended up using machines to cut through the snow. Since then, the site has evolved and researchers continue to trigger and monitor avalanches. Other sudden changes in the cryosphere can also occur through glacial collapse, glacial lake outbursts and landslides. Ice can scour the sea floor and break pipes and cables if they are not buried sufficiently deep, and these are expensive to repair.

There are many ecosystems that depend on the state of the cryosphere. The impacts of climate change on polar bears is often documented in the media: loss of sea ice and hunting opportunities mean that these creatures either starve, drown or spend more time on land, with greater risk to human life. Figure 1.4 shows a polar bear making a surprise appearance at the Churchill Northern Studies Centre, Churchill, Canada.

For land mammals, snow cover provides insulation from cold air temperatures and camouflage. Small mammals dig tunnels beneath the snow. However, without snow covering the land, white Arctic hares and foxes will be a shining beacon to predators. Trees are also dependent on the cryosphere: in the American Northwest,

Figure 1.4. Polar bear peeking around the corner at the Churchill Northern Studies Center, Churchill, Canada. Photo reproduced with kind permission from Arvids Silis.

vegetation growth is limited by the availability of soil moisture so trees tend to follow topographic features and grow in areas where snow drifts form, as shown in figure 1.5. This creates a feedback effect, as the vegetation traps blowing snow leading to more ground accumulation and a greater supply of water for the trees. In the Swedish Arctic, it is said that the trees 'walk' across the Arctic, due to preferential growth and decay linked to snow patterns (Pauker and Seastedt 1996). However, the mechanisms for treeline movement are complicated, with Rundqvist *et al* (2011) identifying vegetation differences consistent with climate change and Van Bogaert *et al* (2011) finding only a small link with climate change.

1.3 Techniques for observing changes

The cryosphere is continually changing on daily to millennial timescales. For example, you may see snow undergo melt-freeze cycles, where it melts during the afternoon and refreezes at night with reduced energy input from the Sun. Large masses of cold ice may not melt, but will slowly move over time, and surface conditions change with precipitation or wind redistribution events. Sea ice is much more dynamic than land ice due to the influence of the ocean circulation. Overall the areal extent of the cryosphere, snow depth and sea ice thickness is declining with recent changes in climate (e.g. Derksen and Brown 2012, Eayrs *et al* 2021, Rackow *et al* 2022, Kacimi and Kwok 2022). In February 2022, Antarctic sea ice extent was the lowest ever observed.

The techniques used to observe the cryosphere depend on the time span of interest and how large an area needs to be observed. Ground-based observations can give

Figure 1.5. Tree growth in Reynolds Creek Experimental Watershed, Idaho, United States follows snow drift patterns.

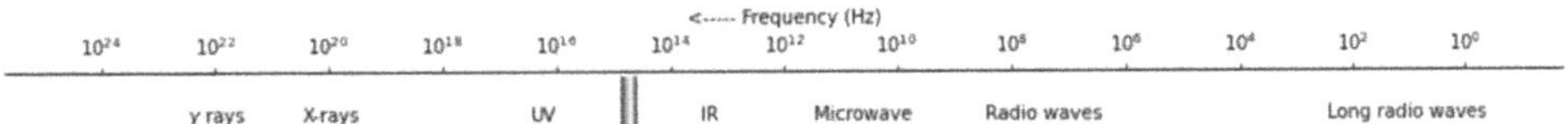

Figure 1.6. Electromagnetic spectrum. Frequency is increasing to the left whereas wavelength is increasing to the right. Visible light wavelength ranges from 0.38 microns (violet) to 0.750 microns (red). Microwave has a longer wavelength (1 mm to 1 m) and lower frequency compared with visible light.

observations within a small area, which can be considered to be a point measurement. Deployment of a network of sensors or repeated measurements at different locations can be used to build a picture of larger spatial coverage. Examples of this would be the SNOTEL snow water equivalent measurement network, or sea ice thickness buoys (Richter-Menge *et al* 2006). Airborne sensors allow measurements at catchment or regional scales. The Airborne Snow Observatories Inc. monitors snowpack water content for water supply management in the United States. For a global perspective, however, satellites are needed.

All remote-sensing techniques used to monitor the cryosphere use some portion of the electromagnetic spectrum, which is shown in figure 1.6. Gamma rays sit at one end of the spectrum, with very short wavelength and high frequency. Gamma rays have been used to monitor the mass of the snowpack, canopy interception losses and snow density (Kodama *et al* 1979, Glynn *et al* 1988, Lundberg 1993, Montgomery *et al* 2018). X-rays come next on the spectrum with slightly longer wavelengths. X-ray tomography is used to measure the 3D microstructural properties of snow and sea ice, including brine and oil pollution (Coléou *et al* 2001, Kerbrat *et al* 2008, Heggli *et al* 2011, Kawamura 1988, Lieb-Lappen *et al* 2017, Desmond *et al* 2021). X-ray diffraction can provide information on the fabric of ice and provide information on the stress history in ice cores and therefore ice sheet mechanics (Montagnat *et al* 2003, Weikusat *et al* 2011).

Between x-ray radiation and the blue side of visible light is ultraviolet radiation, which is not commonly used to observe the cryosphere. Visible light is extremely useful for observing cryosphere extent, as discussed below. After red visible light comes infrared (IR). Near-IR is used to measure the snow specific surface area, i.e. the size of snow crystals and impurities within snow (Arnaud *et al* 2011, Kokhanovsky and Lamare *et al* 2018), whereas longer-wavelength IR, also known as thermal radiation, gives information on the surface temperature and can be used to distinguish between different surface types, e.g. debris on glaciers or leads (open water cracks) in sea ice (Brenning *et al* 2012, Williamson *et al* 2014, Alden *et al* 2018, Reiser *et al* 2020). Microwave radiation is used to sense inside snow and ice, as described below, with longer wavelengths able to penetrate more deeply. Beyond microwave radiation, at the lowest frequency and longest wavelength are radio waves, used for more than half a century to infer layering within ice and the basal properties of glaciers and ice sheets (Robin *et al* 1969, Jordan *et al* 2016, Popov 2020).

The longest time series of Earth observations recorded by satellite uses optical (visible) radiation and is from Landsat, first launched in 1972. Optical imagery is useful because the stark contrast between the whiteness of snow and ice cover and the darker colours of, for example, sea or land surface means that the extent of the cryosphere can be mapped. The downside, however, is that this technique relies on reflected sunlight and therefore cannot be used at night or under cloudy conditions. In addition, clouds also appear optically as bright white and, as such, one of the main challenges in using optical data to monitor the extent of snow is how to distinguish between snow and clouds. For this, thermal bands or a time series of observations (cloud movement) are used, although mixed pixel retrievals remain a challenge (Stillinger *et al* 2019, Richiardi *et al* 2021).

Satellite bands refer to measurement within a range of specific frequencies. Table 1.2 shows the bands of Landsat 9 compared with the original Landsat. Even with the earliest Landsat, the separate bands allow, for example, detection of vegetation, as chlorophyll absorbs radiation in the red portion of the spectrum, i.e. 0.6–0.7 microns (Landsat band 5). This sensitivity (as well as the sensitivity to near-IR) is the basis for the normalised difference vegetation index, or NDVI, algorithm. Advances in instrument design and retrieval algorithms have improved the resolution of the measurements from 60 m to 30 m for visible bands, although it is ultimately limited by the sensor type and wavelength of observation.

There is a lot of physics used to interpret satellite observations in terms of the parameters we wish to know. Corrections are made to geo-register the observations, adjust for calibration of the sensors and account for the effect of the atmosphere (e.g. Doucette *et al* 2013, König *et al* 2019, Li *et al* 2019b, Takenaka *et al* 2020, Kokhanovsky *et al* 2020), which are not described here. Once these corrections have been carried out, the remote-sensing data can be processed to relate the observations

Table 1.2. Bands used for the original and latest Landsat satellites. Adapted from USGS.

Satellite	Band number	Wavelength (microns)	Resolution (m)
Landsat	4	0.5–0.6	60
Landsat	5	0.6–0.7	60
Landsat	6	0.7–0.8	60
Landsat	7	0.8–1.1	60
Landsat 9	1	0.43–0.45	30
Landsat 9	2	0.45–0.51	30
Landsat 9	3	0.53–0.59	30
Landsat 9	4	0.64–0.67	30
Landsat 9	5	0.85–0.88	30
Landsat 9	6	1.57–1.65	30
Landsat 9	7	2.11–2.29	30
Landsat 9	8	0.50–0.68	15
Landsat 9	9	1.36–1.38	30
Landsat 9	10	10.6–11.19	100
Landsat 9	11	11.50–12.51	100

to the desired geophysical properties of the cryosphere. Pope *et al* (2014) provides a summary of how different types of remote-sensing data are used for a wide range of cryospheric applications.

As described above, optical data can be used to map the extent of features such as glaciers (Pfeffer *et al* 2014, Rabatel *et al* 2016), snow (Dumont and Gascoin 2016) or frozen lakes (Barbieux *et al* 2018). However, optical images taken from multiple angles can be used to reconstruct the 3D surface using stereophotogrammetry or structure from motion (Marti *et al* 2016, Chudley *et al* 2019, Anderson *et al* 2019). An example of this applied over a sea ice surface is shown in figure 1.7. Reference target plates, used to stitch and provide a scale for the images, are visible. Repeat measurements under cloud-free conditions mean that volumetric changes, e.g. loss due to melt, can be determined from the time series of observations (Vargo *et al* 2017, Mallalieu *et al* 2017, Li *et al* 2019a, Gunn *et al* 2021, Holmlund 2021). Mass loss can then be calculated if the density is known (mass = density × volume). This can be difficult to estimate, however, as density usually varies both horizontally and vertically.

Microwave frequencies, unlike optical, can be used to sense inside the snow and ice. An additional benefit is that microwave radiation can be used at night and can generally see through clouds, but with the drawback of lower resolution (larger footprint size), which can be of the order of kilometres. How deeply the radiation penetrates the snow and ice depends on the frequency used and microstructure of the snow/ice. Radiation with a wavelength of 20 cm will interact very differently with a spherical ice particle of 1 mm radius than radiation with a wavelength below 1 mm, as illustrated in figure 1.8. How closely packed the particles are, i.e. their density, also matters for determining how radiation propagates through the snow and ice.

Figure 1.7. 3D point cloud of sea ice surface via structure from motion. Photo courtesy of Julien Meloche.

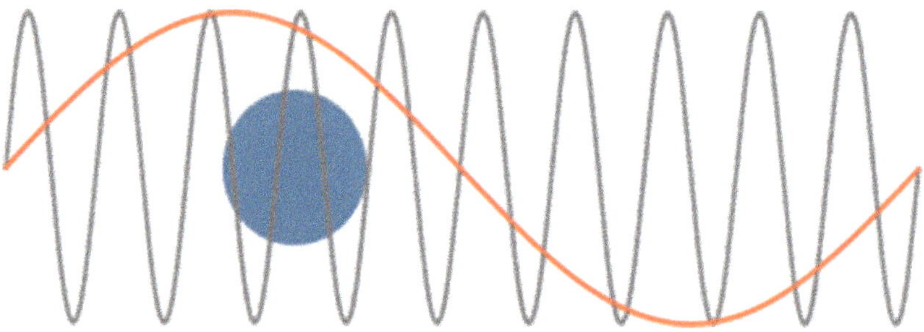

Figure 1.8. Illustration of spherical particle and electromagnetic radiation with wavelength larger (red) and smaller (grey) than the particle radius.

Microwave frequencies are sometimes referred to by bands, but unlike optical sensors these are fixed ranges and can be useful for describing broad electromagnetic behaviour in different portions of the microwave spectrum. Band designations up to 40 GHz are given in table 1.3: these frequencies are most commonly used to gain information about the cryosphere. Higher frequencies penetrate less deeply into the snow and ice, but may still be sensitive to surface properties up to around 243 GHz depending on the water content of the atmosphere.

L-band frequencies are used to monitor soil moisture as it penetrates a few centimetres into the soil. Absorption and scattering of microwave radiation by the liquid water means that the wetter the soil, the less the signal received by the sensor. L-band radiation is relatively insensitive to dry snow, although it can be used to infer information about the snow density at the base of the snowpack (Lemmetyinen *et al* 2016) due to refraction and impedance-matching effects. As with all microwave radiation, the sensitivity to liquid water means that a time series of the L-band is useful for detecting melt events (e.g. Roy *et al* 2015, Xu *et al* 2016); for instance, it has been used to derive a weekly freeze-thaw dataset (Prince *et al* 2018).

As frequency increases, scattering of microwave radiation becomes more sensitive to structural changes within the medium. Figure 1.9 illustrates different scattering

Table 1.3. Microwave frequency band designation up to 40 GHz. Frequency F can be converted to wavelength λ and vice versa through the equation $c = F\lambda$, where c is the speed of light ($\sim 3 \times 10^8$ m s^{-1} in free space).

Band	Frequency range (GHz)	Wavelength
P	0.3–1	0.3–1 m
L	1–2	15–30 cm
S	2–4	7.5–15 cm
C	4–8	3.75–7.5 cm
X	8–12	25–37.5 mm
Ku	12–18	16.7–25 mm
K	18–26	11.5–16.7 mm
Ka	26–40	7.5–11.5 mm

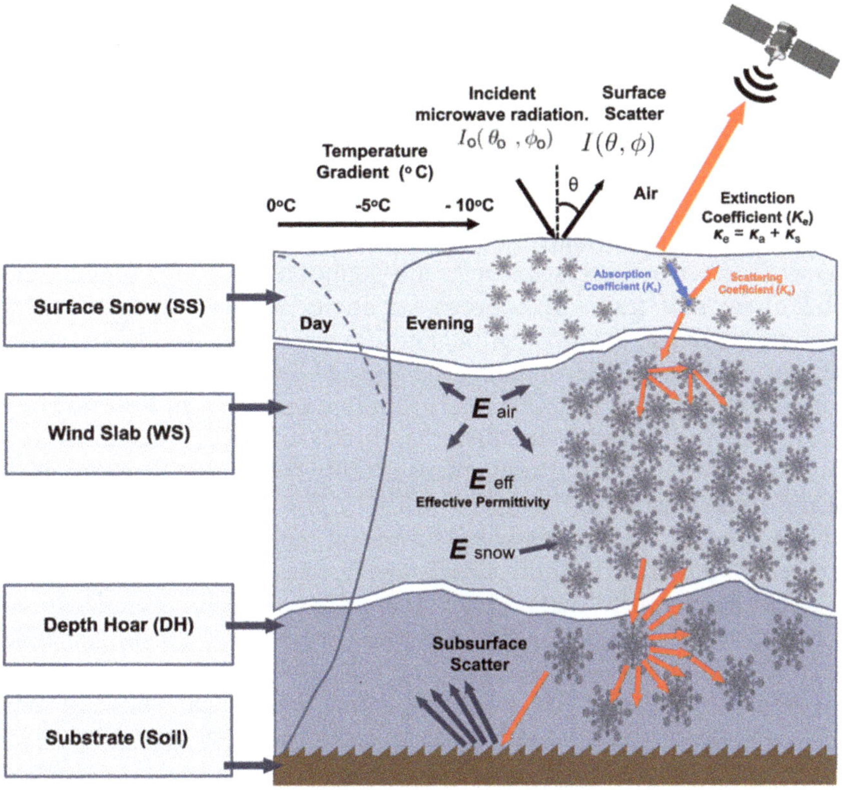

Figure 1.9. Scattering processes within snow. Here, a simple three-layer snowpack is shown above a soil substrate, with increasing size of snow crystals from the surface to the base. Examples of these types of snow crystals are shown in chapter 4, figure 4.4. Radiation is scattered from individual crystals and at boundaries between layers. A diurnal change in temperature gradient is also indicated, with warmer surface temperatures during day and cooling in the evening. Image courtesy of Harry Clayton.

processes within snow, i.e. scattering from the surface, the interface between snow and soil and within the snow layers themselves. Radiation is also absorbed and emitted within the layers. Microwave radiation is scattered at any interface between materials with different dielectric properties (i.e. the way in which electromagnetic waves interact with it), e.g. brine pockets in sea ice or air bubbles within pure ice. The more scattering material there is, the more the signal is reduced. It is this reduction in signal that enables remote sensing of snow mass, provided the other properties of the snow, such as stratigraphy and the size of the microstructure, are known.

Passive sensors measure electromagnetic radiation that is naturally emitted (e.g. microwave and thermal) or reflected (visible light from the Sun) whereas *active* sensors emit a signal and measure the return. Scattering and absorption processes are common to both sensor types and can be accounted for with a radiative transfer equation, described in chapter 2. A third sensor type is an *altimeter*, which is an active sensor that measures the echo as a function of time. The time taken for the signal to return to the sensor can then be converted to the distance between the sensor and reflecting surface, given the speed of light. This can be used to build a digital elevation model (Slater *et al* 2018) relative to a reference height or to measure sea ice thickness (Quartly *et al* 2019).

Although scattering of C-band radiation is weakly dependent on snow microstructure, there is enough sensitivity in temporal changes to allow retrievals of snow depth in mountainous areas, which have typically been challenging to observe (Lievens *et al* 2022). C-band data have also been used to identify glacier facies (zones of glaciers with different properties; Thakur *et al* 2017), monitoring of avalanches (Yang *et al* 2020) and permafrost deformation along the Qinghai–Tibet Railway (Chen *et al* 2012). Ka-band frequencies are commonly used for remote sensing of snow, but there is a trade-off between sensitivity to mass and how deeply the radiation reaches inside the snow. At higher frequencies the signal may saturate and it is then not possible to determine the thickness of deeper snowpacks (Mätzler and Wiesmann 1999, Grody 2008). One further technique allows the flow rate of glaciers to be monitored and relies on identification of backscatter patterns in radar images, e.g. from crevassed regions (Pritchard *et al* 2005), and tracking the changes in locations of these over time.

There are many remote-sensing sensors available, and ever-evolving techniques and tools to better interpret data for understanding the cryosphere. Here the broad behaviour has been briefly outlined; for a more comprehensive explanation of the physics behind remote sensing of the cryosphere, see Tedesco (2014) and Emery and Camps (2017). The time series of satellite observations provides information on changes in the cryosphere over the last five decades or so, but to understand how the climate has changed over millennia we need to look at ice cores and other deposition records.

1.4 Deposition climate records

As snow densifies to ice, air becomes trapped within the ice matrix and eventually becomes isolated within individual bubbles (described in more detail in chapters 4 and 5). The atmosphere at the time the pores completely close is preserved.

Extraction of the gas contained within the bubbles enables analysis of CO_2, methane and other gas concentrations over the last 800 000 years (Lüthi *et al* 2008). To do this, sections of ice are extracted and melted to extract the gas, as illustrated by videos from the Niels Bohr Institute and Royal Society. Isotopes in the meltwater can also be analysed to provide a temperature record (Jouzel and Masson-Delmotte 2007): colder conditions mean fewer heavy isotopes such as ^{18}O and deuterium transported towards the poles (Masson-Delmotte *et al* 2005). The relative difference between deuterium and ^{18}O has also been related to sea ice concentration (Kurosaki *et al* 2020). Sea ice extent can also be determined from methanesulphonic acid within the ice core, related to algal blooms governed by sea ice extent (Abram *et al* 2010). Ice cores contain a wealth of minerals and other chemical constituents that give information about formation and transport processes, e.g. dust storm origin and changes in circulation patterns (Legrand and Mayewski 1997, Brook and Buizert 2018).

Dating of ice core chemical analysis can use an understanding of accumulation rates and densification processes: how quickly the pores close and how quickly layers are compressed and buried within the ice. Other methods of dating include identifying and counting distinct layers within the ice cores (Hammer *et al* 1978, Lemieux-Dudon *et al* 2010). Spatial differences in accumulation rates and ice flow conditions can complicate the dating process, but cores from different locations can be compared by identifying markers such as volcanic layers in the cores (Vinther *et al* 2006), as illustrated in figure 1.10. A good understanding of all physical and

Figure 1.10. Ice core from West Antarctic ice sheet with dark volcanic layer. Copyright © Heidi Roop, 'Ice Core', Attribution-ShareAlike 2.0 Generic (https://www.flickr.com/photos/oregonstateuniversity/45690255762).

chemical processes involved in geochemical changes within the cores gives greater confidence in the dating of the ice cores.

Ice cores are not the only way to determine past climate history, though. Preserved remnants of plant and animal material relate to precipitation and temperature conditions of the climate at the time and can be dated through radiocarbon or isotope analysis, thermoluminescence or counting of annual layers (Walker 2005). Peat bogs, for example, contain plant and pollen fossils. High levels of sphagnum moss microfossils can be indicative of cooler and/or wetter climates whereas well-decomposed peat with evidence of trees suggests warmer, drier climates (Blackford 2000). The particular species of *Sphagna* matter, however, as some are more suited to drier conditions and are thus indicative of these (Magnan *et al* 2018). Diatoms, microscopic silicified algae (Likens 2009; see figure 1.11 for an image of one species) within salt marsh sediments can give an indication of sea level rise as different species compete more favourably depending on their exposure along the slope down to the sea (Gehrels 2000, Woodroffe and Long 2009).

Layered sediment deposition in lakes can be used to derive the history of freeze/thaw lake conditions. When frozen, fine particles are deposited out of the water, whereas coarser sediment deposited into the lake forms summer layers (Walker 2005). Coral reefs and shells within marine sediment deposits provide useful information on sea surface and near-surface temperature conditions and atmospheric and ocean circulation patterns, with sub-annual resolution due to shell growth processes (Wanamaker *et al* 2011). Positions of shells within glacial till can be used to infer growth and retreat of ice extent (England 1999). Speleothems,

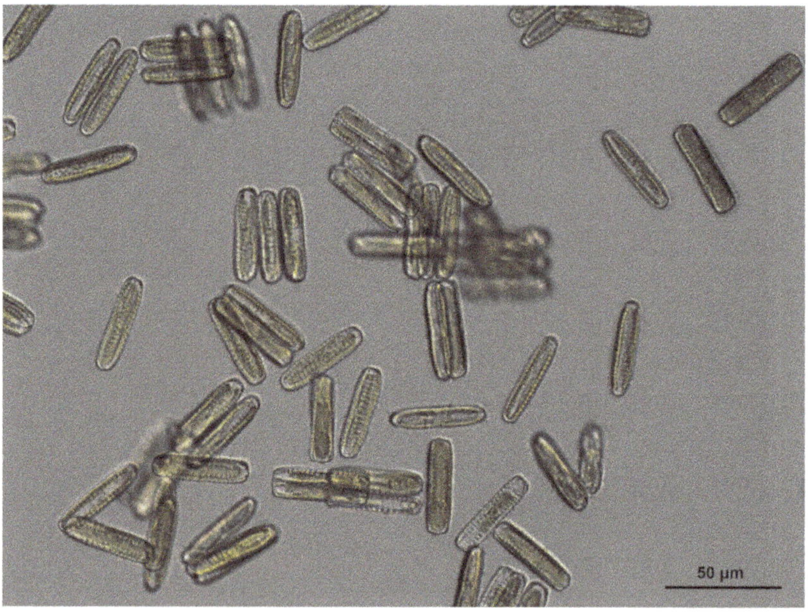

Figure 1.11. *Pinnularia borealis*, one of the diatom species used by Woodroffe and Long (2009) to reconstruct a record of relative sea level rise. Copyright BCCM/DCG (https://bccm.belspo.be/about-us/bccm-dcg).

e.g. stalactites and stalagmites, also have a layered structure, related to changes in the availability of water as they are formed. A hiatus in growth can indicate the past presence of permafrost above that prevented water flow to the caves below, giving a record of past permafrost extent (Vaks and Gutareva *et al* 2013). Conversely, past speleothem growth in areas with permafrost today suggests previous permafrost thaw. Under warming climate conditions, loss of sea ice may increase the rate of permafrost thaw due to enhanced sensible and latent heat transported inland from the ocean (Vaks and Mason *et al* 2020).

Dendrochronology links the growth patterns of trees to the climate. Cellular growth during the spring and summer differs, leading to the formation of tree rings. The rate of growth, or thickness of rings, provides information on past climate (Walker 2005). Dendrochronology has also been used to demonstrate that thermokarst expansion in the latter half of the last century was due to precipitation changes rather than air temperature (Agafonov *et al* 2004). Trees grow differently when they are inclined to the vertical, giving an estimate of the date of thermokarst formation, i.e. collapse of soil due to permafrost thaw. The age of trees growing on lateral or terminal glacial moraines has also provided information on past glacial extents (Leclercq *et al* 2014).

1.5 Climate tipping points and the high latitudes

The 5th IPCC report has identified several 'tipping points' related to climate change (IPCC 2013). Once a tipping point is reached, irreversible changes happen to a system in equilibrium. Numerous studies focus on dynamical systems to study the resilience of a system and how many equilibrium states it can survive on.

Taking as an example figure 1.12, we can imagine the Earth in a state of equilibrium. Each equilibrium state can be thought about as a position in a sill. If the system is perturbed, in our case, i.e. with a strong raise in air temperature, the system

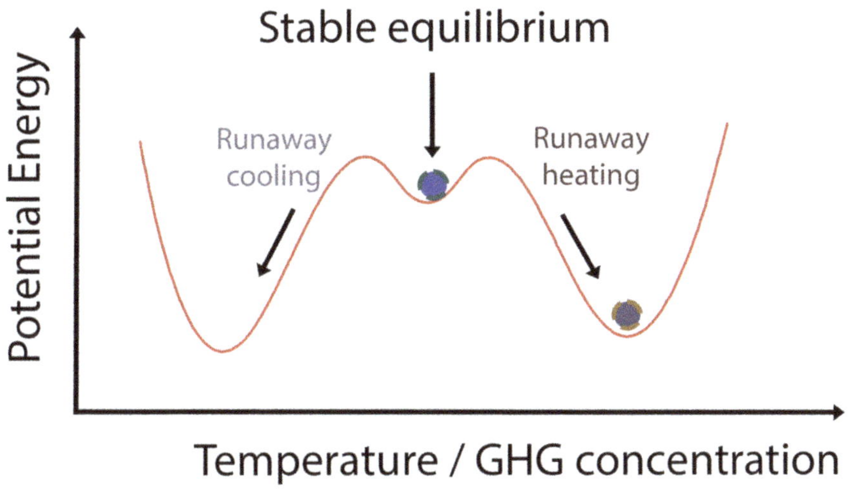

Figure 1.12. Dynamical systems, adapted from Harrison (2013).

'moves' to a state of unstable equilibrium; and if the perturbation is strong enough or applied for a sufficient amount of time, the system will fall into a new state of equilibrium. The Earth system itself will adjust to the new equilibrium, but the human species might not be able to do so, therefore, from our point of view, the new equilibrium will be worse than the current one. The critical processes identified by the IPCC that can reach a tipping point in most current scenarios are as follows (figure 1.13):

1. Shutdown/slowdown of the Atlantic Meridional Overturning Circulation.
2. West Antarctic ice sheet disintegration.
3. Greenland ice sheet disintegration.
4. Amazon rainforest dieback.
5. Permafrost and methane hydrates.
6. Coral reef die-off.
7. West African monsoon shift.
8. The El Niño–Southern Oscillation.
9. Arctic sea ice melt.

From this list, we can gather the importance of the high-latitude areas and why there is a pressing need to understand how they work and further how to mitigate the changes we observe today. In fact, five out of nine points of this list that can have strong repercussions on the whole Earth system relate to processes happening at the poles.

In this primer we shall describe some of the mechanisms mentioned above, highlighting their links to the global climate in chapters 3, 5, 6 and 7. Interpretation of climate records requires an understanding of the physical processes that govern

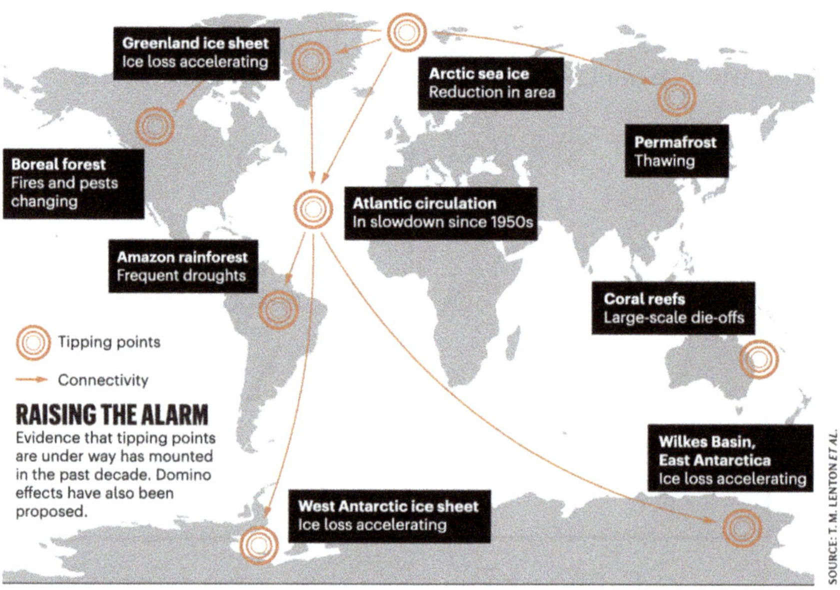

Figure 1.13. Climate tipping points. Reprinted/adapted by permission from Springer: *Nature*, 'Climate tipping points too risky to bet against', by Lenton *et al* (2019).

their formation and quantify the uncertainty in dating methods. The ability to model components of the cryosphere to understand how it may change with future climate warming also relies on a good representation of the physical processes involved. There are many similarities in behaviour between permafrost, snow, land and sea ice because they are underpinned by basic physics and the physics of ice in particular. The next chapter provides the physical basis of this behaviour, which is then applied in subsequent chapters on each component of the cryosphere.

References

Abram N J *et al* 2010 Ice core evidence for a 20th century decline of sea ice in the Bellingshausen Sea, Antarctica *J. Geophys. Res. Atmos.* **115** D23101

Adolph A, Albert M R and Hall D K 2018 Near-surface temperature inversion during summer at Summit, Greenland, and its relation to MODIS-derived surface temperatures *Cryosphere* **12** 907–20

Agafonov L, Strunk H and Nuber T 2004 Thermokarst dynamics in Western Siberia: insights from dendrochronological research *Palaeogeogr. Palaeoclimatol. Palaeoecol* **209** 183–96

Anderson K, Westoby M J and James M R 2019 Low-budget topographic surveying comes of age: structure from motion photogrammetry in geography and the geosciences *Prog. Phys. Geogr.: Earth Environ.* **43** 163–73

Arnaud L *et al* 2011 Measurement of vertical profiles of snow specific surface area with a 1 cm resolution using infrared reflectance: instrument description and validation *J. Glaciol.* **57** 17–29

Barbieux K, Charitsi A and Merminod B 2018 Icy lakes extraction and water-ice classification using Landsat 8 OLI multispectral data *Int. J. Remote Sens.* **39** 3646–78

Barnett T P, Adam J C and Lettenmaier D P 2005 Potential impacts of a warming climate on water availability in snow-dominated regions *Nature* **438** 303–9

Blackford J 2000 Palaeoclimatic records from peat bogs *Trends Ecol. Evol.* **15** 193–8

Brenning A *et al* 2012 Thermal remote sensing of ice-debris landforms using ASTER: an example from the Chilean Andes *Cryosphere* **6** 367–82

Brodzik M J and Stewart J S 2016 Near-real-time SSM/I-SSMIS EASE-grid daily global ice concentration and snow extent, Version 5 *Technical Report* NASA

Brook E J and Buizert C 2018 Antarctic and global climate history viewed from ice cores *Nature* **558** 200–8 ISSN: 1476-4687

Brown J *et al* 2002 Circum-Arctic Map of Permafrost and Ground-ice Conditions, Version 2 *Data Sheet* GGD318 National Snow and Ice Data Center

Chen F *et al* 2012 Interaction between permafrost and infrastructure along the Qinghai-Tibet Railway detected via jointly analysis of C- and L-band small baseline SAR interferometry *Remote Sens. Environ.* **123** 532–40

Chudley T R *et al* 2019 High-accuracy UAV photogrammetry of ice sheet dynamics with no ground control *Cryosphere* **13** 955–68

Coléou C *et al* 2001 Three-dimensional snow images by X-ray microtomography *Ann. Glaciol.* **32** 75–81

G Consortium 2005 GLIMS Glacier Database, Version 1 *NASA Technical Report* National Snow and Ice Data Center

Derksen C and Brown R 2012 Spring snow cover extent reductions in the 2008–2012 period exceeding climate model projections *Geophys. Res. Lett.* **39** L19504

Desmond D S *et al* 2021 Investigation into the geometry and distribution of oil inclusions in sea ice using non-destructive X-ray microtomography and its implications for remote sensing and mitigation potential *Mar. Pollut. Bull.* **173** 112996

Doucette P *et al* 2013 Image georegistration methods: a framework for application guidelines. 2013 IEEE Applied Imagery *2013 IEEE Applied Imagery Pattern Recognition Workshop (AIPR)* pp 1–14

Dufour F G U, Bartelt P and Ammann W J 2000 Overview of the 1999 measurements at the SLF test site Vallée de la Sionne *Proc. Int. Snow Science Workshop (Blue Sky, MT, October 1–6)* pp 527–34

Dumont M and Gascoin S 2016 Optical remote sensing of snow cover *Land Surface Remote Sensing in Continental Hydrology* ed B Nicolas and Z Mehrez (Amsterdam: Elsevier) ch 4 pp 115–37

Eayrs C *et al* 2021 2021 Rapid decline in Antarctic sea ice in recent years hints at future change *Nat. Geosci.* **14** 460–4

Emery B and Camps A 2017 *Introduction to Satellite Remote Sensing: Atmosphere* (Elsevier: Amsterdam)

England J 1999 Coalescent Greenland and Innuitian ice during the Last Glacial Maximum: revising the Quaternary of the Canadian High Arctic *Quat. Sci. Rev.* **18** 421–56

Fang Z *et al* 2018 Reduced sea ice protection period increases storm exposure in Kivalina, Alaska *Arctic Sci.* **4** 525–37

Foster J L *et al* 2005 Quantifying the uncertainty in passive microwave snow water equivalent observations *Remote Sens. Environ.* **94** 187–203

Gehrels W R 2000 Using foraminiferal transfer functions to produce high-resolution sea-level records from salt-marsh deposits, Maine, USA *Holocene* **10** 367–76

Glynn J E *et al* 1988 An airborne gamma ray snow survey of a forest covered area with a deep snowpack *Remote Sens. Environ.* **26** 149–60

Grody N 2008 Relationship between snow parameters and microwave satellite measurements: theory compared with advanced microwave sounding unit observations from 23 to 150 GHz *J. Geophys. Res. Atmos.* **113** D22108

Gunn G, Jones B M and Rangel R C 2021 Unmanned aerial vehicle retrieval of snow depth over freshwater lake ice using structure from motion *Front. Remote Sens* **2** 15

Hammer C U *et al* 1978 Dating of Greenland ice cores by flow models, isotopes, volcanic debris, and continental dust *J. Glaciol.* **20** 3–26

Harrison K 2013 Building resilient communities *M/C J.* **16** 1–8

Heggli M *et al* 2011 Measuring snow in 3-D using X-ray tomography: assessment of visualization techniques *Ann. Glaciol.* **52** 231–6

Holmlund E 2021 Aldegondabreen glacier change since 1910 from structure-from-motion photogrammetry of archived terrestrial and aerial photographs: utility of a historic archive to obtain century-scale Svalbard glacier mass losses *J. Glaciol.* **67** 107–16

Hope C and Schaefer K 2016 Economic impacts of carbon dioxide and methane released from thawing permafrost *Nat. Clim. Change* **6** 56–9

IPCC 2013 *Climate Change 2013: The Physical Science Basis. Contribution of Working Group I to the Fifth Assessment Report of the Intergovernmental Panel on Climate Change* eds. T F Stocker, D Qin, G-K Plattner *et al* (Cambridge: Cambridge University Press)

Jordan T M *et al* 2016 An ice-sheet-wide framework for englacial attenuation from ice-penetrating radar data *Cryosphere* **10** 1547–70

Jouzel J and Masson-Delmotte V 2007 EPICA Dome C Ice Core 800KYr deuterium data and temperature estimates *PANGAEA* (https://doi.org/10.1594/PANGAEA.683655)

Kacimi S and Kwok R 2022 Arctic snow depth, ice thickness, and volume from ICESat-2 and CryoSat-2: 2018–2021 *Geophys. Res. Lett.* **49** e2021GL097448

Kawamura T 1988 Observations of the internal structure of sea ice by X-ray computed tomography *J. Geophys. Res. Oceans* **93** 2343–50

Kerbrat M *et al* 2008 Measuring the specific surface area of snow with X-ray tomography and gas adsorption: comparison and implications for surface smoothness *Anglais. Atmos. Chem. Phys.* **8** 1261–75

Key J, Drinkwater M and Ukita J 2007 IGOS Cryosphere Theme Report *WMO/TD*

Kodama M *et al* 1979 An application of cosmic-ray neutron measurements to the determination of the snow-water equivalent *J. Hydrol.* **41** 85–92

Kokhanovsky A *et al* 2018 On the reflectance spectroscopy of snow *Cryosphere* **12** 2371–82

Kokhanovsky A *et al* 2020 The determination of snow albedo from satellite measurements using fast atmospheric correction technique *Remote Sens.* **12** 234

König M, Hieronymi M and Oppelt N 2019 Application of Sentinel-2 MSI in Arctic Research: evaluating the performance of atmospheric correction approaches over Arctic Sea Ice *Front. Earth Sci.* **7** 22

Kurosaki Y *et al* 2020 Reconstruction of sea ice concentration in Northern Baffin Bay using deuterium excess in a coastal ice core from the Northwestern Greenland ice sheet *J. Geophys. Res. Atmos.* **125** e2019JD031668

Leclercq P W *et al* 2014 A data set of worldwide glacier length fluctuations *Cryosphere* **8** 659–72

Legrand M and Mayewski P 1997 Glaciochemistry of polar ice cores: a review *Rev. Geophys.* **35** 219–43

Lemieux-Dudon B *et al* 2010 Consistent dating for Antarctic and Greenland ice cores *Quat. Sci. Rev.* **29** 8–20

Lemmetyinen J *et al* 2016 Snow density and ground permittivity retrieved from L-band radiometry: application to experimental data *Remote Sens. Environ.* **180** 377–91

Lenton T M *et al* 2019 Climate tipping points—too risky to bet against *Nature* **575** 592–5

Li T *et al* 2019a Resolving fine-scale surface features on Polar Sea Ice: a first assessment of UAS photogrammetry without ground control *Remote Sens.* **11** 784

Li X *et al* 2019b The recent developments in cloud removal approaches of MODIS snow cover product *Hydrol. Earth Syst. Sci.* **23** 2401–16

Lieb-Lappen R M, Golden E J and Obbard R W 2017 Metrics for interpreting the microstructure of sea ice using X-ray micro-computed tomography *Cold Reg. Sci. Technol.* **138** 24–35

Lievens H *et al* 2022 2022 Sentinel-1 snow depth retrieval at sub-kilometer resolution over the European Alps *Cryosphere* **16** 159–77

Likens G E 2009 *Encyclopedia of Inland Waters* (New York: Academic)

López-Carr D and Marter-Kenyon J 2015 Human adaptation: manage climate-induced resettlement *Nature* **517** 265–7

Lundberg A 1993 Evaporation of intercepted snow–review of existing and new measurement methods *J. Hydrol.* **151** 267–90

Lüthi D *et al* 2008 CO_2 record from the EPICA Dome C 1999 (EDC99) ice core (Antarctica) covering 650 to 800 kyr BP measured at the University of Bern, Switzerland. Type: data set *PANGAEA* https://doi.pangaea.de/10.1594/PANGAEA.710901

Magnan G *et al* 2018 Impact of the Little Ice Age cooling and 20th century climate change on peatland vegetation dynamics in central and northern Alberta using a multi-proxy approach and high-resolution peat chronologies *Quat. Sci. Rev.* **185** 230–43
Mallalieu J *et al* 2017 An integrated structure-from-motion and time-lapse technique for quantifying ice-margin dynamics *J. Glaciol.* **63** 937–49
Marti R *et al* 2016 Mapping snow depth in open alpine terrain from stereo satellite imagery *Cryosphere* **10** 1361–80
Masson-Delmotte V *et al* 2005 Holocene climatic changes in Greenland: different deuterium excess signals at Greenland Ice Core Project (GRIP) and NorthGRIP *J. Geophys. Res. Atmos.* **110** D14102
Mätzler C and Wiesmann A 1999 Extension of the microwave emission model of layered snowpacks to coarse-grained snow *Remote Sens. Environ.* **70** 317–25
Montagnat M *et al* 2003 Lattice distortion in ice crystals from the Vostok core (Antarctica) revealed by hard X-ray diffraction; implication in the deformation of ice at low stresses *Earth Planet. Sci. Lett.* **214** 369–78
Montgomery L, Koenig L and Alexander P 2018 The SUMup dataset: compiled measurements of surface mass balance components over ice sheets and sea ice with analysis over Greenland *Earth Syst. Sci. Data* **10** 1959–85
Pauker S J and Seastedt T R 1996 Effects of mobile tree islands on soil carbon storage in tundra ecosystems *Ecology* **77** 2563
Pfeffer W T *et al* 2014 The Randolph Glacier inventory: a globally complete inventory of glaciers *J. Glaciol.* **60** 537–52
Pope A *et al* 2014 Open access data in polar and cryospheric remote sensing *Remote Sens* **6** 6183–220
Popov S 2020 Fifty-five years of Russian radio-echo sounding investigations in Antarctica *Ann. Glaciol.* **61** 14–24
Prince M *et al* 2018 Northern Hemisphere surface freeze-thaw product from Aquarius L-band radiometers *Earth Syst. Sci. Data* **10** 2055–67
Pritchard H *et al* 2005 *J. Geophys. Res. Earth Surf.* **110** C03020
Quartly G D *et al* 2019 Retrieving sea level and freeboard in the Arctic: a review of current Radar altimetry methodologies and future perspectives *Remote Sens.* **11** 881
Rabatel A, Dedieu J P and Vincent C 2016 Spatio-temporal changes in glacier-wide mass balance quantified by optical remote sensing on 30 glaciers in the French Alps for the period 1983–2014 *J. Glaciol.* **62** 1153–66
Rackow T *et al* 2022 2022 Delayed Antarctic sea-ice decline in high-resolution climate change simulations *Nat. Commun.* **13** 637
Raup B *et al* 2007 The GLIMS geospatial glacier database: a new tool for studying glacier change *Glob. Planet. Change* **56** 101–10
Reiser F, Willmes S and Heinemann G 2020 A new algorithm for daily sea ice lead identification in the Arctic and Antarctic Winter from thermal-infrared satellite imagery *Remote Sens.* **12** 1957
Richiardi C *et al* 2021 A revised snow cover algorithm to improve discrimination between snow and clouds: a case study in Gran Paradiso National Park *Remote Sens.* **13** 1957
Richter-Menge J A *et al* 2006 Ice mass-balance buoys: a tool for measuring and attributing changes in the thickness of the Arctic sea-ice cover *Ann. Glaciol.* **44** 205–10

Robin G D Q, Evans S and Bailey J T 1969 Interpretation of radio echo sounding in polar ice sheets *Phil. Trans. R. Soc. A* **265** 437–505
Roy A *et al* 2015 Evaluation of spaceborne L-band radiometer measurements for terrestrial freeze/thaw retrievals in Canada *IEEE J. Sel. Top. Appl. Earth Obs. Remote Sens.* **8** 4442–59
Rundqvist S *et al* 2011 Tree and Shrub expansion over the past 34 years at the tree-line near Abisko, Sweden *AMBIO* **40** 683–92
Slater T *et al* 2018 A new digital elevation model of Antarctica derived from CryoSat-2 altimetry *Cryosphere* **12** 1551–62
Stillinger T *et al* 2019 Cloud masking for Landsat 8 and MODIS Terra over snow-covered terrain: error analysis and spectral similarity between snow and cloud *Water Resour. Res.* **55** 6169–84
Sturm M, Goldstein M A and Parr C 2017 Water and life from snow: a trillion dollar science question *Water Resour. Res.* **53** 3534–44
Takenaka H *et al* 2020 Geolocation correction for geostationary satellite observations by a phase-only correlation method using a visible channel *Remote Sens.* **12** 2472
Tedesco M 2014 *Remote Sensing of the Cryosphere* (Wiley: New York)
Thakur P K *et al* 2017 Estimation of snow cover area, snow physical properties and glacier classification in parts of Western Himalayas using C-Band SAR data *J Indian Soc Remote Sens* **45** 525–39
Vaks A *et al* 2013 Speleothems reveal 500,000-year history of Siberian permafrost *Science* **340** 183–6
Vaks A *et al* 2020 Palaeoclimate evidence of vulnerable permafrost during times of low sea ice *Nature* **577** 221–5
Van Bogaert R *et al* 2011 A century of tree line changes in sub-Arctic Sweden shows local and regional variability and only a minor influence of 20th century climate warming *J. Biogeogr.* **38** 907–21
Vargo L J *et al* 2017 Using structure from motion photogrammetry to measure past glacier changes from historic aerial photographs *J. Glaciol.* **63** 1105–18
Vaughan D G *et al* 2013 Climate change 2013: the physical science basis *Contribution of Working Group I to the Fifth Assessment Report of the Intergovernmental Panel on Climate Change* Observations: Cryosphere (Geneva: IPCC) pp 317–82
Vinther B M *et al* 2006 A synchronized dating of three Greenland ice cores throughout the Holocene *J. Geophys. Res. Atmos.* **111** D006921
Walker M 2005 *Quaternary Dating Methods* (New York: Wiley)
Wanamaker A D, Hetzinger S and Halfar J 2011 Reconstructing mid- to high-latitude marine climate and ocean variability using bivalves, coralline algae, and marine sediment cores from the Northern Hemisphere *Palaeogeogr. Palaeoclimatol. Palaeoecol* **302** 1–9
Weikusat I *et al* 2011 Subgrain boundaries in Antarctic ice quantified by X-ray Laue diffraction *J. Glaciol.* **57** 111–20
Williamson S N *et al* 2014 Estimating temperature fields from MODIS land surface temperature and air temperature observations in a Sub-Arctic alpine environment *Remote Sens* **6** 946–63
Woodroffe S A and Long A J 2009 Salt marshes as archives of recent relative sea level change in West Greenland *Quat. Sci. Rev.* **28** 1750–61
Xu X *et al* 2016 Freeze/Thaw detection and validation using Aquarius' L-Band backscattering data *IEEE J. Sel. Top. Appl. Earth Obs. Remote Sens.* **9** 1370–81
Yang J *et al* 2020 Automatic detection of regional snow avalanches with scattering and interference of C-band SAR data *Remote Sens.* **12** 2781

Chapter 2

Basic physics

2.1 Thermodynamics of the cryosphere

One of the most important impacts of the cryosphere on the climate is on the albedo of the Earth's surface: this is the proportion of incoming sunlight reflected at the surface. This is given by equation (2.1):

$$\alpha = \frac{S^{\uparrow}}{S^{\downarrow}}, \tag{2.1}$$

where α is the albedo, $S^{\uparrow}$ is the upwelling (reflected) solar radiation and $S^{\downarrow}$ is the downwelling (incoming) solar radiation. From this, the amount of solar radiation absorbed by the surface, S_{net}, is

$$S_{\mathrm{net}} = (1 - \alpha)S^{\downarrow}. \tag{2.2}$$

The lower the albedo, the more energy is absorbed by the surface. Approximate values for the albedo of common natural features of the Earth's surface are given in table 2.1. These values are typical, but there is quite a bit of variation due to factors such as the roughness (or conversely how flat and smooth a surface is), the chemical constituents or impurities, species of plant, water content of the soil, and, for snow, the size of the snow crystals. Figure 2.1 illustrates a change in albedo for soil: a section of the soil has been recently ploughed, bringing moisture to the surface. The wetter soil has a lower albedo, so reflects less sunlight and appears darker to the eye.

The impact of albedo on the energy input to the planet can be illustrated with the values in table 2.1. For a given energy input of 800 W m^{-2} from the Sun, the amount absorbed by soil would be 640 W m^{-2}, whereas if the soil was covered with fresh snow, the amount of solar radiation absorbed would only be 120 W m^{-2}. Similarly, for sea and sea ice, the difference is large. The problem is that under a warming climate the cryosphere shrinks, resulting in lower albedos and more energy absorbed by the planet. This is known as a *feedback loop*, where warming causes physical changes that lead to more warming, and is one of the reasons why the cryosphere is so important.

doi:10.1088/978-0-7503-3647-5ch2

Table 2.1. Approximate albedo values for some natural materials.

Material	Approximate albedo [unitless]
Soil	0.2
Fresh snow	0.85
Trees	0.15
Sea water	0.05
Sea ice	0.6

Figure 2.1. Soil with recently ploughed section appears darker because it has a higher moisture content. Photo reproduced with kind permission from Ian Davenport.

Another important equation governing cryospheric behaviour in a warming climate is the Clausius–Clapeyron equation: this determines the amount of moisture the air can hold as a function of its temperature. The Clapeyron equation relates changes in pressure P, temperature T and change in volume ΔV as through the specific entropy of phase change, ΔS, as

$$\frac{dP}{dT} = \frac{\Delta S}{\Delta V}. \tag{2.3}$$

For exchanges with the gaseous phase, the ideal gas law can be substituted into the Clapeyron equation to form the Clausius–Clapeyron equation:

$$\frac{de_v}{dT} = \frac{Le_v}{RT^2}, \tag{2.4}$$

where e_v is the saturation vapour pressure, R is the universal gas constant and L is the specific latent heat (also known as enthalpy) of evaporation if phase changes between liquid and gas are considered, or the specific heat of sublimation if phase change takes place between solid and gas phases.

Neglecting the small temperature dependence of the latent heat, this equation can be integrated to form

$$\left[\log_e e_v\right]_{e_v^0}^{e_v^f} = -\frac{L}{R}\left[\frac{1}{T}\right]_{T^0}^{T^f}, \tag{2.5}$$

where e_v^0 and e_v^f are the initial and final values of the vapour pressure, and T^0 and T^f are the initial and final values of the temperature, giving a final equation:

$$e_v^f = e_v^0 \exp\left\{\frac{L}{R}\left[\frac{1}{T^0} - \frac{1}{T^f}\right]\right\}. \tag{2.6}$$

From equation (2.6), it can be seen that for higher temperatures the vapour pressure is higher, i.e. the air will hold more moisture. It is this property that governs precipitation around the globe: water is picked up from warmer areas, moved around by wind (in turn linked to pressure differences) and deposited in colder regions.

Thus the impacts of climate change on the cryosphere are complicated. Broadly, a warming climate causes more melt, with an additional warming feedback through the albedo changes, but also may bring more precipitation. This could be snow or rain, depending on the temperature. Snow is good news for the cryosphere: more solar radiation is reflected and less heat is absorbed. Rain is very bad news for the cryosphere as its high energy content causes rapid melt or thaw of frozen surfaces, and potential flooding. Clouds themselves shield the surface from solar radiation but emit and contribute to thermal radiation absorbed by the surface. Whether this warms or cools the surface depends on the energy balance, and we use conservation equations to understand changes in the cryosphere.

2.2 Conservation equations

The physics of the cryosphere used to predict the behaviour of cryospheric components relies on two main principles: conservation of mass and conservation of energy. Water is not destroyed but is converted from one form to another, between ice, liquid and vapour depending on the energy balance. Similarly, energy is converted from one form to another and may result in a local warming or cooling of the cryosphere. Basic equations are presented here: these will crop up again in different forms throughout the book, though are essentially the same thing but with different assumptions. Not all terms may be included (e.g. the energy balance equation excludes the kinetic energy of precipitation because it is small). Here, they will be applied to a generic 'block of cryosphere'. This could be a block of glacial ice, a section of snowpack or a cuboid of sea ice.

The generic conservation equation in integral form for a volume dV is

$$\frac{d}{dt}\int_V \gamma_k \Psi dV = -\sum_k \int_S \mathbf{J}\cdot d\mathbf{S} + \int_V \sigma dV, \tag{2.7}$$

or, alternatively, the change in quantity Ψ (e.g. temperature) is equal to the sum of the fluxes across the boundaries (first term on the right-hand side) and sources/sinks (second term on the the right-hand side). In equation (2.7), t is time, V is the volume and γ_k is the partial density of constituent k, where k is air, water vapour, liquid water or ice. Ψ is the quantity being conserved, $\mathbf{J}$ is the flux across the boundary enclosing boundary $\mathbf{S}$ and σ is the source/sink density.

Treatment of the first term depends on whether the grid is fixed (Eulerian) or moving with the material (Lagrangian) as

$$\frac{d}{dt} = \frac{\partial}{\partial t} + \frac{\partial z}{\partial t}\frac{\partial}{\partial z}. \tag{2.8}$$

For a Lagrangian grid whose boundaries move but contains largely the same material (apart from fluxes across the boundaries and sources/sinks) then the second term in equation (2.8) drops out and the derivative with respect to time in the conservation equation (2.7) can be replaced with the partial derivative.

The partial density γ_k of a material is a key concept of mixture theory—where different phases of the same material are present. Snow is a good example as it is a mixture of ice, liquid water, vapour and air. γ_k is the mass of that material per unit volume of the whole mixture. Therefore the partial density of the ice phase of snow is substantially lower than the intrinsic density of ice (mass of a constituent per volume of that constituent), which is 916.7 kg m^{-3} at 0 °C.

Thus to keep track of the ice mass, the equation might look like this:

$$\frac{\partial}{\partial t}\int_V \gamma_i dV = -\int_S \mathbf{U}_i \cdot d\mathbf{S} + \sum_k \int_V M_{i,k} dV\left(1 - \delta_{i,k}\right), \tag{2.9}$$

where i denotes ice, $\mathbf{U}_i$ is the ice mass flux into the system and $M_{i,k}$ is the mass rate of change from phase k to ice. δ is the Kronecker delta function so there is no contribution if $k = i$ (i.e. ice to ice). Equation (2.9) translates to change in ice mass = snowfall + condensation from vapour + refreeze − melt − sublimation. Another flux term would be needed if it is raining. If the amount of liquid water is important for a particular application (e.g. for surface runoff or infiltration) then a separate mass conservation equation for the liquid component might be considered. The mass balance equation can also be applied to the entire cryospheric block, making no distinction between ice, liquid and vapour phases. It is, after all, all water. In this case, there is no third term in the equation, and it is only flow through the surface and through the base of the material that contribute to mass changes.

In principle, mass flux input is easy to determine: go out and measure it, or use numerical weather prediction outputs. For now, we will pretend that these give you perfect values. Determining phase change rates, though, is harder, and that is where the energy balance equations come in.

A simple energy balance equation for the surface might look like this:

$$Q = S^{\downarrow} - S^{\uparrow} + L^{\downarrow} - L^{\uparrow} + H + LE + M + G, \tag{2.10}$$

where Q is the change in energy, which will result in either phase change or a temperature change. S and L are solar (shortwave) and thermal (longwave) radiation. The up and down arrows denote the direction of radiative flux. Downward solar and thermal radiation can be measured with radiometers (a pyranometer for solar and a pyrgeometer for thermal). In the case of solar radiation, also known as shortwave radiation, the upward radiation is purely due to reflection of the downwelling radiation. The proportion of solar radiation reflected is given by the albedo, as described earlier in equation (2.1). Thermal radiation, also termed longwave due to its longer wavelength, has a small reflected component, governed by the thermal emissivity ϵ. For ice, thermal emissivity is around 0.99, so only about 1% of downwelling thermal radiation is reflected at the surface, and is often assumed to be negligible in models. The surface also emits thermal radiation, so the bulk of the upwelling thermal radiation will be emission, governed by the Stefan–Boltzmann equation:

$$L^{\uparrow} = \epsilon \sigma T^4, \tag{2.11}$$

where T is the surface temperature and σ is the Stefan–Boltzmann constant (5.67×10^{-8} W m^{-2} K^{-4}).

Returning to the energy balance equation (2.10), H denotes sensible heat transfer; this is convection by another name. LE is latent heat transfer, i.e. evaporation/sublimation/condensation processes. Both H and LE are dependent on the wind, temperature and humidity profile of the atmosphere above the surface. There are ways of measuring them (Martin and Lejeune 1998), but it can be particularly tricky in cold regions as the amount of vapour in the air is smaller compared to warmer regions (see earlier Clausius–Clapeyron; equation (2.6)) and the wind profile can be a lot smoother. There are also ways of modelling them (Morris 1989): this is rather complicated and, depending on the application, may not be important, so has not been described here.

There are two terms left in the energy balance equation. M is advected heat: the heat transferred from rain or snow precipitation at a different temperature T_p to the surface T_i, where c is the specific heat capacity and $\rho \Delta z$ is the precipitation mass added:

$$M = c\rho\Delta z\left(T_p - T_i\right). \tag{2.12}$$

If the precipitation is at the same temperature as the surface then $M = 0$. G is the ground heat flux, the conduction of heat from the surface underneath the block. This can be measured or modelled with a simple heat diffusion equation in 1D:

$$G = K\frac{\partial T}{\partial z}. \tag{2.13}$$

The change in energy can result in a change in temperature for a unit volume:

$$Q = c\gamma\Delta T, \tag{2.14}$$

and/or a phase change:

$$Q = \gamma L_k. \tag{2.15}$$

Some cryospheric components move rapidly, e.g. icebergs or melt water, so a conservation of momentum would be needed to determine rate of flow. An example of this is given in chapter 6.

2.3 Radiative transfer theory

Conservation of energy principles are applied to interpret satellite data, as discussed in the previous chapter. The time-independent radiative transfer equation governs the change in radiation (or satellite signal) in a specific direction (see figure 1.9) and is given in Picard *et al* (2018) as:

$$\begin{aligned}\mu\frac{\partial \mathbf{I}(\mu, \phi, z)}{\partial z} &= -\boldsymbol{\kappa}_e(\mu, \phi, z)\mathbf{I}(\mu, \phi, z)\\ &+ \frac{1}{4\pi}\int\!\!\int_{4\pi} \mathbf{P}(\mu, \phi, \mu', \phi', z)\,\mathbf{I}(\mu', \phi', z)d\Omega'\\ &+ \boldsymbol{\kappa}_a(\mu, \phi, z)\alpha T\mathbf{1}.\end{aligned} \tag{2.16}$$

On the left-hand side of the equation is the change in reduced specific intensity $\mathbf{I}$ with height z along the azimuthal direction ϕ and μ, which is the cosine of the incidence angle. The first term on the right-hand side are the losses due to absorption and from scattering away from the direction of (ϕ, μ). The extinction coefficient $\boldsymbol{\kappa}_e$ that governs losses is a sum of the absorption coefficient $\boldsymbol{\kappa}_a$ and scattering coefficient $\boldsymbol{\kappa}_s$. The second term on the right-hand side accounts for radiation scattered from other directions into (ϕ, μ) and is given by the integration of the phase function $\mathbf{P}$ over the solid angle Ω'. The final term in the equation is the source term: this is the radiation emitted by the material itself as a function of temperature T. The speed of light in the medium is accounted for through α, and $\mathbf{1}$ is a unit vector (equal emission in all directions). For altimeter applications, where the echo is measured as a function of time, a time-dependent radiative transfer equation is needed (Larue *et al* 2021).

In addition to the radiative transfer equation, boundary conditions are needed to describe the reflection and transmission across boundaries, e.g. snow–air, ice–air or between different layers within the snow. For flat interfaces, the Fresnel equations can be used, but other formulations are needed for rough interfaces (not described here). Much of the complexity in equation (2.16) is in how to represent the scattering and absorption coefficients, and the phase function that represents the probability of radiation being scattered from one direction to another, which is beyond the scope of this primer. The microstructure of the scatterers and density of the material are key parameters, and their evolution under different temperature and pressure conditions is a function of the ice properties and the exchanges between solid, liquid and gaseous forms of water.

2.4 Properties of ice

The physical properties of ice at microscales govern the macroscale behaviour of ice bodies. Even though the cryosphere is composed of different elements, they often react in the same general way to changes in the local environmental conditions. Water is such a unique material and governs the world around us, and its behaviour is all due to its molecular structure.

Ice is the crystalline form of water, H_2O, formed at temperatures below 0 °C; it is colourless, transparent and its crystal lattice is of a hexagonal system. The molecular crystalline structure of ice has four hydrogen atoms linked to each oxygen atom: two by hydrogen bonds (the water molecule), and two by polar covalent links, as illustrated in figure 2.2. The bond angles are near tetrahedral. Ice can form different crystalline structures depending on the temperature and pressure conditions (see Hobbs 1974), but the form we experience is 1 h, as shown in figure 2.2(b).

Water is fairly unique in that it is one of the only materials to exist in vapour, liquid and solid at a temperature and pressure that we commonly experience. Ice is an unusual material in that the solid phase is less dense than liquid: it expands when it freezes and so icebergs float. From the change in volume term in the Clapeyron equation (2.3), this leads to a negative pressure–temperature gradient, as shown in the phase diagram in figure 2.3, inset. A rise in pressure results in a drop in freezing point, termed freezing-point depression. As the pressure increases, ice will transition to liquid water at lower temperatures. In the sea this phenomenon is generally negligible: an external pressure of 100 atm (around 1000 m depth), would decrease the ice melting point by 1 °C. However, at higher pressures the freezing-point depression may allow pressure melting to occur, as shown in figure 2.4 and in a time lapse video demonstrating the process of regelation. It is this property that can result in melting at the base of ice bodies, known as basal melting, and rapid flow of ice bodies such as glaciers, as described in chapter 5. The presence of impurities such as

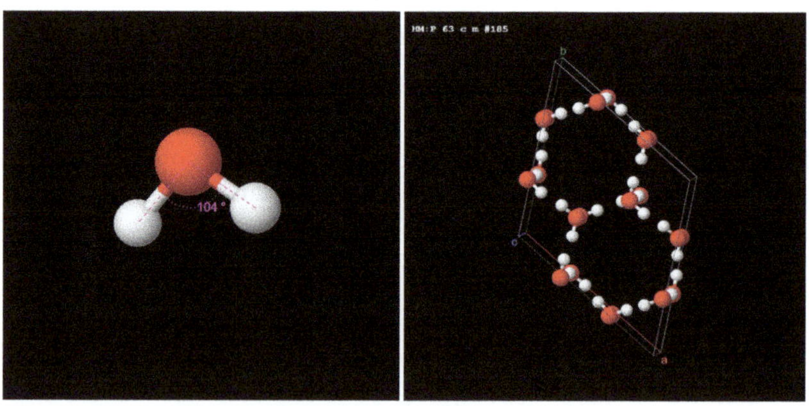

Figure 2.2. Molecular structure of ice drawn with https://molview.org. Left: Single molecule (oxygen atom shown in red, hydrogen atoms shown in white). Right: Hexagonal form of ice 1 h structure.

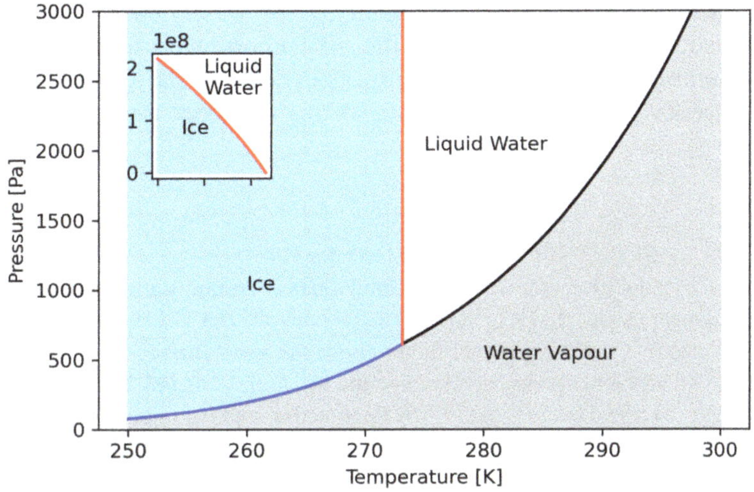

Figure 2.3. Simple illustration of ice phase diagram. Solid, liquid and gas phases coexist at the triple point. Division between solid and gaseous phases and between liquid and gaseous phases are given by the Clausius–Clapeyron equation (2.6), whereas the division between solid and liquid was given by the approximation in Wagner *et al* 2011 (their equation (6)).

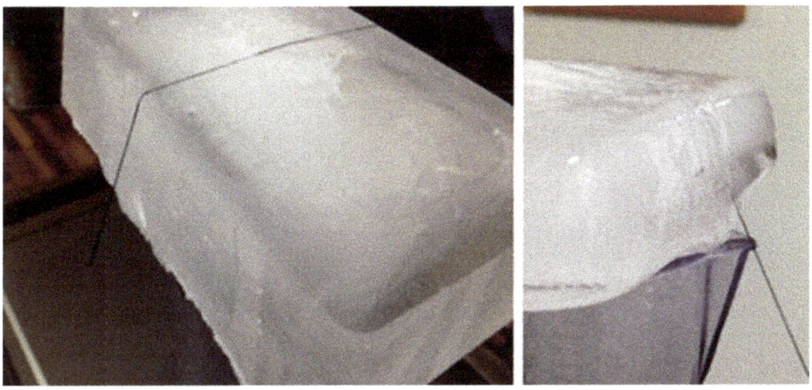

Figure 2.4. Regelation of ice. A wire or thread with suspended weights placed across a block of ice will slowly cut through the ice by pressure melting, and will resolidify once the pressure has been removed, i.e. above the wire, to leave a solid block. (a) Time at 0 hr, (b) time at approximately 2 hr.

salt, particularly important for sea ice, can also lower the freezing point, as described in chapter 6.

At a temperature of 0 °C the density of ice is 916.7 kg m^{-3}, indicating that it floats on water. Technically, there is enough interstitial space in the crystal lattice for a water molecule, although this is not thought to occur commonly. As temperature decreases, the density of the ice lattice increases to 934.0 kg m^{-3} at −180°. Being less dense than liquid water, ice expands as it forms and can be responsible for some of the worst weathering of natural and man-made materials. Changes between the liquid and ice phase can have a large impact on the structure of the landscape because of the change in volume occupied by the water when it changes phase, leading to distinct features unique to the cryosphere. The following chapter on permafrost describes how freeze-thaw processes lead to the formation of pingos, palsas, retrogressive thaw slumps and the sorting of stones into circles, stripes and polygons.

References

Hobbs P V 1974 *Ice Physics* vol 1 (Oxford: Clarendon)

Larue F *et al* 2021 Radar altimeter waveform simulations in Antarctica with the snow microwave radiative transfer model (SMRT) *Remote Sens. Environ.* **263** 112534

Martin E and Lejeune Y 1998 Turbulent fluxes above the snow surface *Ann. Glaciol.* **26** 179–83

Morris E M 1989 Turbulent transfer over snow and ice *J. Hydrol.* **105** 205–23

Picard G, Sandells M and Löwe H 2018 SMRT: an active–passive microwave radiative transfer model for snow with multiple microstructure and scattering formulations (v1.0) *Geosci. Model Dev.* **11** 2763–88

Wagner W *et al* 2011 New equations for the sublimation pressure and melting pressure of H_2O ice Ih *J. Phys. Chem. Ref. Data* **40** 043103

IOP Publishing

Melody Sandells and Daniela Flocco

Chapter 3

Permafrost and periglacial environments

Periglacial environments refers to those affected by frost processes, whereas permafrost is a term to describe ground that has been at a temperature of 0 °C or below for two or more consecutive years, regardless of whether the ground is frozen, unfrozen or partially frozen. Continuous cold conditions can be a function of latitude or elevation, leading to a distinction between Arctic (latitudinal) and mountain (altitudinal) permafrost (Dobinski 2011). Although commonly thought of as applying to land, this can also refer to the seabed, a relic from when sea levels were ~130 m lower than today (Barry and Gan 2011). Permafrost may not be completely frozen and may contain water if the temperature is at the triple point or if the melting point is depressed due to pressure or the presence of salt, e.g. in marine permafrost (Dobinski 2011). In this chapter we briefly examine the physics of the freeze-thaw of frozen soil, describe the characteristics of the permafrost, look at the mechanism for the formation of stone polygons and discuss the implications of permafrost thaw for society.

3.1 Frost heave

Periglacial processes that govern the evolution of the landscape arise from freeze-thaw expansion and contraction, flows of water and water-saturated soil and wind erosion (French 1996). At a granular level, the physics of freezing is very interesting. Water itself expands upon freezing, but soil containing moisture may or may not show changes on freezing depending on the properties of the soil. Frost heave, where the land surface rises as it freezes, depends on the size and shape of the soil particles, saturation of the soil and rate of cooling. Taber (1930) identified this and demonstrated experimentally that under certain conditions uplift does not occur with freezing as the water is pushed downwards and laterally, whereas under other conditions the water may be drawn up from other areas. In these experiments, distinct ice layers formed within freezing clay from water pulled up through capillary pressure, as illustrated in figure 3.1(b). Pore freezing, downward propagation of the

doi:10.1088/978-0-7503-3647-5ch3

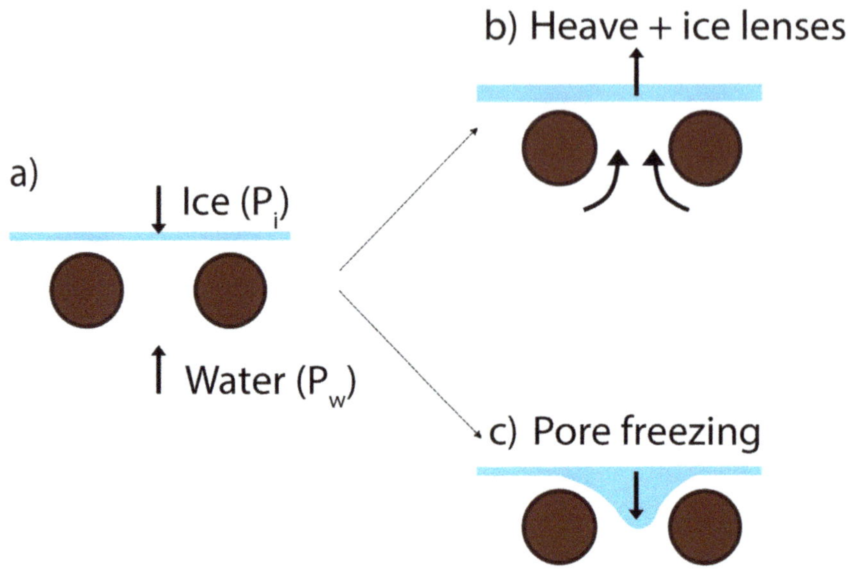

Figure 3.1. Frost-heave processes. Freezing can cause water migration towards the ice and lead to distinct ice lenses (segregated ice) as shown in (b), or the freezing may penetrate into the pore space as in (c), depending on pore size and relative pressure of the ice and water. After French (1996).

freezing front as illustrated in figure 3.1(c), occurs if the pressure difference between ice and water ($P_i - P_w$) is greater than the threshold pressure ΔP_t given by the ice–water interface energy, γ_{iw}, and effective radius of the pore, r_p, according to equation (3.1) (Peppin and Style 2013):

$$\Delta P_t = \frac{2\gamma_{iw}}{r_p} \tag{3.1}$$

If the pressure from the water is greater than ice, then water moves up through the soil leading to segregation: formation of ice lenses. From the Clapeyron equation that relates pressure, volume and temperature changes (see figure 2.3 and equation (2.3)), this pressure difference threshold can then be related to temperature, i.e. the temperature at which frost heave occurs for a particular geometry can be derived, given by equation (3.2), where T_p is the temperature at which pore freezing rather than segregation occurs, T_f is the freezing temperature, ρ_w is the density of water and L_f is the latent heat of freezing:

$$T_p = T_f\left(1 - \frac{2\gamma_{iw}}{\rho_w L_f r_p}\right). \tag{3.2}$$

Consequently, frost heave and ice segregation tends to occur for fine-grained material, whereas pore freezing occurs in coarse-grained material such as sand (French 1996). Permafrost often contains large sections of solid ice, as shown in figure 3.2. This can be a large problem under warming conditions as the volume

Figure 3.2. Section of ice within permafrost on Herschel Island, Canada. Photo by Boris Radosavljevic, taken on the Yukon 2012 Expedition by the COPER group of the Alfred Wegener Institute in Potsdam, Germany. Image redistributed under the CC BY 2.0 licence.

occupied by the ice is greater than the volume of liquid water, leading to subsidence as the permafrost thaws[1]. Thawing of the permafrost can lead to building collapse (as shown in figure 3.3), reduced slope stability and destruction of roads, railways and pipework (Anisimov *et al* 2007, Key *et al* 2007). Permafrost prevents release of greenhouse gases, so if the permafrost thaws the atmospheric effect leads to warming and further thawing. Since the early 1980s, there has been a general increase in permafrost temperatures by up to 2 °C and increase in the depth of annual thaw and refreeze by up to 90 cm (Vaughan *et al* 2013).

3.2 Temperature of the permafrost

Permafrost is a remnant of previous glacial periods, including the Little Ice Age from around 400 years up to 150 years ago. The oldest permafrost, however, has been dated to 650 000 years ago and was found in Eastern Siberia (Murton *et al* 2022). At a certain depth, the ground must be warmer than 0 °C due to heat from the Earth's core. Permafrost has a finite thickness (up to around 1.5 km), but varies spatially. Different parts of the permafrost can be identified from its temperature profile, as shown in figure 3.4. At the surface, the active layer describes ground that thaws and refreezes annually. The thickness of the active layer can be monitored with a frost probe, forced through the soil until the top of the solid permafrost is found. Immediately below the active layer, the ground experiences fluctuations in the temperature but remains below zero. Under this, the isothermal permafrost has a

[1] *Melt* refers to the phase change of the entire material, whereas a mixed material containing solids *thaws* under warming conditions. Permafrost, like a frozen chicken, does not melt!

Figure 3.3. Collapsed house due to permafrost. Photo reproduced by kind permission from Peter Webley.

constant temperature gradient to the base of the permafrost, which ends where the temperature exceeds 0 °C.

Changes in the permafrost temperature, T, may be modelled with the 1D heat diffusion equation (Ling and Zhang 2004, Nicolsky *et al* 2009):

$$C\frac{\partial T}{\partial t} = \frac{\partial}{\partial z}\left(k\frac{\partial T}{\partial z}\right), \tag{3.3}$$

where the effective volumetric heat capacity C incorporates the volumetric heat capacity C_v and phase changes of the volumetric unfrozen water content θ (L is the latent heat of fusion):

$$C = C_v + L\frac{\partial \theta}{\partial T}. \tag{3.4}$$

In permafrost, it can be assumed there is no liquid water transport (Ling and Zhang 2004). Although there is unfrozen water, it is held within the soil matrix. With this assumption, volumetric unfrozen water content can be modelled as a function of the temperature:

$$\theta = \frac{\rho_b}{\rho_u}a|T - 273.15|^b, \tag{3.5}$$

where ρ_b is the bulk density of soil, ρ_u is the density of liquid water and a and b are constants dependent on soil type.

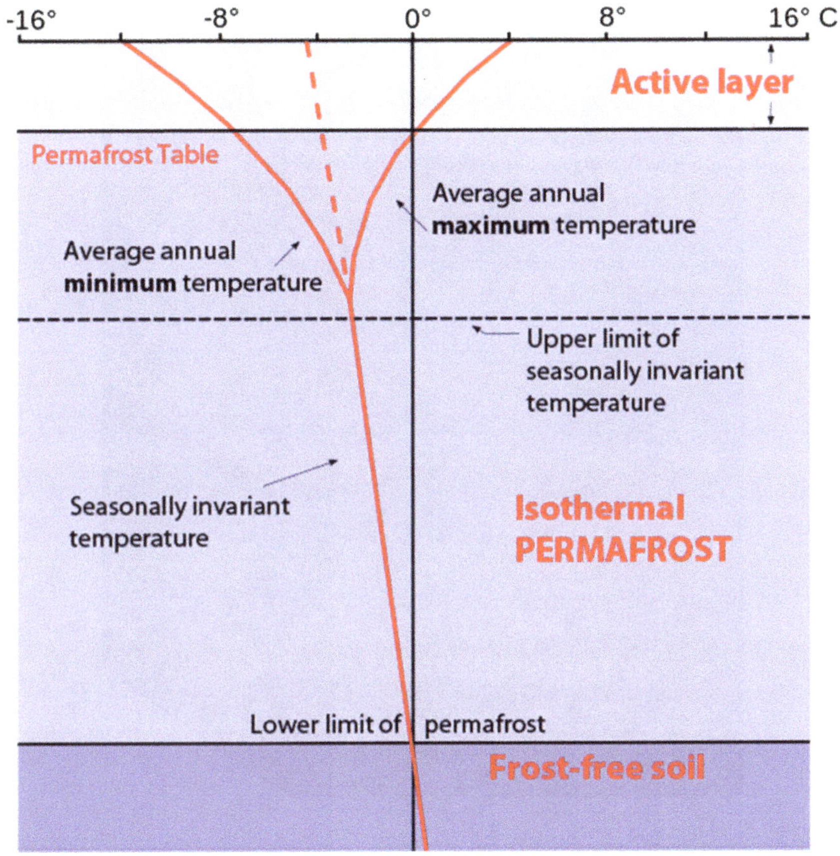

Figure 3.4. Typical temperature profile of permafrost. Image by HylgeriaK, redistributed under the CC0 1.0 universal licence.

Unfrozen sections of soil within the permafrost are known as *taliks*, shown in figure 3.5. They may be just under the active layer (*supra-permafrost* talik) or within the permafrost (*intra-permafrost* talik). The unfrozen ground underneath the permafrost is also termed the *sub-permafrost* talik. Taliks may form beneath unfrozen lakes, and in shallow permafrost may extend to the base of the permafrost (known as *through* taliks).

Taliks usually exist because of the contact between ground and unfrozen water features: lakes, rivers or ground water. A *cryopeg* is another type of unfrozen soil feature, but occurs in regions with high dissolved salt content, which lowers the freezing point of water (Barry and Gan 2011). As discussed in chapter 6, the molar concentration of dissolved solute C_m and the type of solute, quantified by the cryoscopic constant k_f, govern the lowering of the freezing point ΔT according to

$$\Delta T = k_f C_m. \tag{3.6}$$

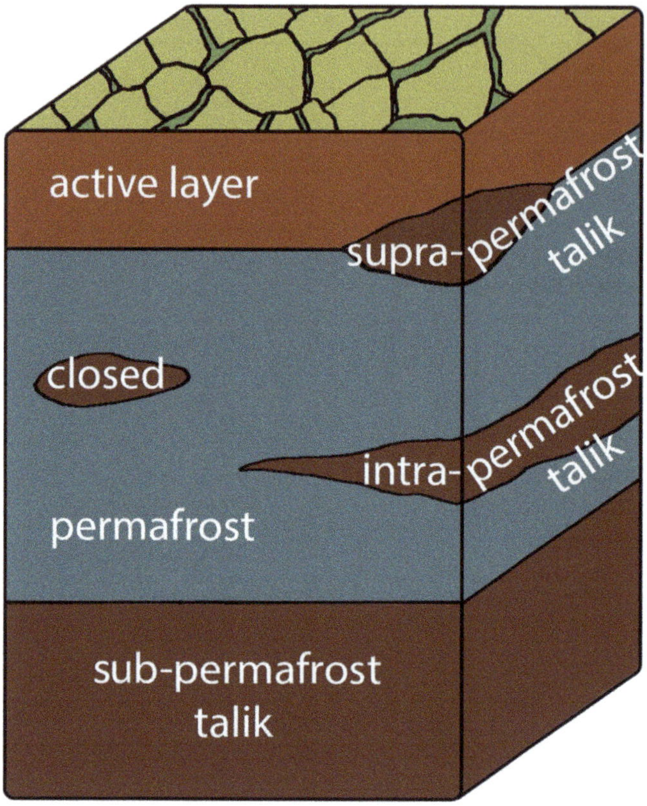

Figure 3.5. Illustration of unfrozen sections of soil within the permafrost, known as taliks.

3.3 Periglacial landscape formation

Freezing and thawing of the active layer causes changes in the land surface. For example, pingos are conical or dome-shaped features in the land surface, similar to Piercy Pingo, shown in the top of figure 3.6. The bottom panel shows a pingo that is just starting to form. Lateral pressure forces talik water upwards, which then freezes and expands to form an ice core at the heart of the pingo. Growth over time means that pingos can become large, up to 50 m high and 300 m in diameter (Barry and Gan 2011), and can rupture when the ice core melts.

On a smaller scale (1–8 m high and 5–25 m wide) though formed by similar processes are palsas, which consist of alternating layers of ice and soil or peat (Barry and Gan 2011). Palsas are typically grouped together and are formed in discontinuous permafrost. Figure 3.7 illustrates a collection of palsas. By experimentally removing the snow, Seppälä (1982) showed that palsa formation depended on snow depth.

Ice polygons, as shown in figure 3.8, start to form in winter. Dropping temperatures cause shrinking and therefore cracking in the frozen active layer. In spring the cracks are filled with liquid water, which then freeze in the following winter. Expansion due to phase change from liquid to ice widens the crack. This process

Figure 3.6. Top: Piercy Pingo on the North Slope of Alaska. Bottom: A pingo is starting to form. Images reproduced by kind permission from Ronald Daanen.

depends on the temperature and soil type, and can continue for many years. In cold temperatures, ice may remain in the cracks and form ice wedges.

Melting of the ice contained within the permafrost can lead to collapse in the land surface. The land surface slumps down and leaves behind depressions in the land surface, known as thermokarst. Batagaika Crater in Siberia, as captured by Landsat and shown in figure 3.9, is an example of a megaslump. Craters that subsequently fill with water are known as thermokarst lakes. Coastal erosion in permafrost areas can lead to slope failure, or exposure of ice that subsequently ablates, i.e. is removed by melt or sublimation processes (Lantuit and Pollard 2008). Collapse of the land

Figure 3.7. Aerial view of palsas. Photo by Dentren, released under the Creative Commons Attribution-Share Alike 3.0 Unported license.

Figure 3.8. Ice wedge polygons, North Slope, Alaska. Image reproduced by kind permission from Ronald Daanen.

surface under these circumstances is known as retrogressive thaw slump. Herschel Island, Canada, has experienced an increase in the number and total area of retrogressive thaw slumps over the last 50 years (Lantuit and Pollard 2008).

A general overview of Arctic glacial landforms is shown in this video. Small-scale changes in the land surface, such as the movement of individual stones, can lead to the formation of much larger-scale patterns in the land surface.

Figure 3.9. Batagaika Crater, imaged by Landsat and provided by NASA.

3.4 Sorting of stones

Expansion and contraction with soil freezing and thawing processes moves ice, the soil and anything contained within the soil such as stones, rocks and boulders. Redistribution of stones leads to the formation of surface patterns such as polygons, stripes or circles (figure 3.10). Kessler and Werner (2003) developed a numerical model to explain the reasons for formation of these different patterns by the same mechanism. Two factors govern the patterns formed by sorting of stones, which were applied to an initially random grid of modelled stones and allowed to move over hundreds of freeze-thaw cycles.

Firstly, the freezing front penetrates more rapidly in areas with higher stone densities due to the lower water content. Expansion upon freezing of the soil water applies a force normal to the freezing front. Stones are pushed upwards but then roll down the local gradient. This process is modelled with

$$\delta \mathbf{x}_{\mathrm{ls}} = K_{\mathrm{ls}} \nabla S, \tag{3.7}$$

where $\delta \mathbf{x}_{\mathrm{ls}}$ is the distance the stone is moved downslope according to the local surface S, which is related to stone concentration. K_{ls} is a diffusion constant that governs how quickly the stones move.

The second effect is lateral squeezing of stone-filled grooves and regions, which pushes the stones upwards. Stones may roll downslope along the line of compression or spill out to the side. Similar to equation (3.7), this process is represented by

$$\delta \mathbf{x}_{\mathrm{sq}} = \left| K_{\mathrm{sq}} \nabla U \right| \hat{\mathbf{u}}, \tag{3.8}$$

where $\delta \mathbf{x}_{\mathrm{sq}}$ is the distance parallel to the line of compression, K_{sq} is a diffusion parameter governing the rate of stone transport, U is surface uplift from squeezing of

Figure 3.10. Top: Non-sorted polygons, Ellef Ringnes Island, Canada. Middle: Sorted stripes from Ny-Ålesund, Svalbard. Bottom: Sorted circles from Kvadehukken, Svalbard. Photos reproduced with kind permission from Ina Timling.

the stones, and $\hat{\mathbf{u}}$ is a weighted unit vector determined from the axis of compression and a randomly oriented vector. The relative weighting determines the degree to which the stones are confined along the line of compression. By varying the degree of confinement, the mean surface slope and the initial stone concentration, Kessler and Werner (2003) were able to reproduce stone patterns of circles, islands, polygons and stripes, indicating that permafrost properties could be inferred by the stone patterns formed.

3.5 Patterned ground on Mars

Patterned ground has even been found on Mars (see figure 3.11). This implies the presence of water and freeze-thaw processes (Paepe *et al* 2001, Barrett *et al* 2017). Mars is much colder than Earth, though, with an average temperature of 220 K (Jakosky and Roger 2001), cold enough for CO_2 to precipitate and cover the surface with solid CO_2 in winter. Penetration of solar radiation can warm the sub-surface in spring, leading to sublimation and the emission of gaseous CO_2, along with dust via jets that may close and reopen. The direction of the dark deposits indicate the wind direction during emission (Aye *et al* 2019).

Clasts, or rock fragments, have been found to form ordered structures similar to patterns of sorted stones on Earth. An example of striped patterning of clasts on the slope of a crater on Mars is shown in figure 3.11, right. Barrett *et al* (2017) investigated different mechanisms of clast movement such as boulder ratcheting from adjacent fracture formation and closure processes due to CO_2 frost, and also gravitational movement down slopes. Whilst they rejected these two mechanisms as responsible for the observed clast distributions, they suggested periglacial processes could cause the observed patterns despite Martian temperatures thought to be too cold for freeze-thaw cycles. Other remote-sensing images suggest features similar to the Earth's cryosphere such as glaciers and thermokarst slumps (Lucchitta 1981, Hubbard *et al* 2014).

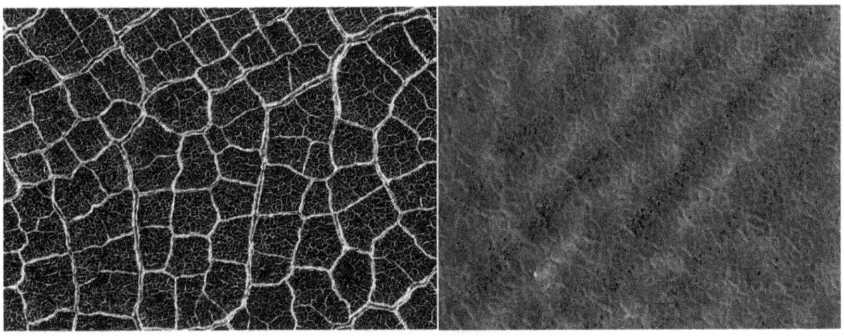

Figure 3.11. Left: Patterned ground found on the surface of Mars. Courtesy of NASA/JPL-Caltech/University of Arizona. Right: Clast stripes in a Martian crater, HiRISE image ESP 017685 2490, courtesy NASA/JPL-Caltech/UArizona.

3.6 Thawing of permafrost

Thawing of the permafrost is a particular concern because of damage to infrastructure and release of methane. Four million people will be affected by the likely damage to infrastructure by 2050 (Hjort and Karjalainen 2018). Structures at risk include roads, buildings, railways, bridges, pipelines and airstrips. Potential mitigation strategies include methods to extract heat during winter (e.g. thermosyphons), reduce heat absorption in the summer (e.g. change surface albedo) and/or improve soil stability (e.g. controlled thaw or reinforcement). The costs of these are in the region of tens of billions of U.S. dollars (Hjort and Streletskiy 2022). Hope and Schaefer (2016) estimate the economic costs of carbon dioxide and methane release from permafrost thaw by 2200 will be anywhere in the range of 3 to 166 trillion U.S. dollars.

Permafrost traps carbon within a frozen matrix and inhibits drainage, thereby keeping the soil within the active layer wetter. With permafrost thaw, oxygen becomes more readily available to microbes within the soil and the consequent degradation of organic matter results in increased release of carbon dioxide. However, anerobic bacteria more suited to wetter conditions release methane in addition to carbon dioxide (Schuur and Abbott 2011, Herndon 2018, Knoblauch *et al* 2018). Methane is a far more damaging greenhouse gas as it has a stronger effect and operates over shorter timescales. Figure 3.12 illustrates the spatial variability in emissions and how methane emission from wetter areas are more damaging than the carbon dioxide emissions from drier areas.

Release of biological specimens frozen for thousands of years could change the ecology or expose us to unknown pathogens. A viable giant virus with similar structure and behaviour to large pathogenic viruses was discovered in 30 000-year-old Siberian permafrost (Legendre *et al* 2014). Human and animal disease outbreaks in the Russian Arctic include tick-borne encephalitis, rabies, leptospirosis, brucellosis, tularaemia and anthrax (see box) (Revich *et al* 2012). Not only will these be more prevalent with warming, human risk of exposure to these will increase with more time spent outdoors.

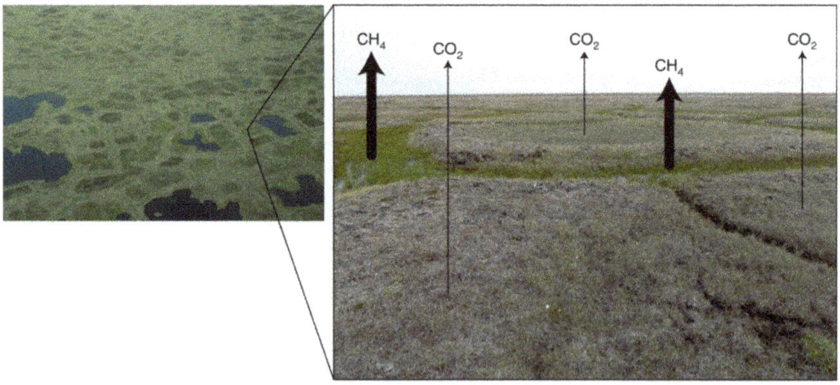

Figure 3.12. Spatial distribution in greenhouse gas emission from thawing permafrost. Drier areas predominantly emit CO_2, whereas wetter areas emit both CO_2 and CH_4 (methane). The size of the arrows denotes relative potential contribution to warming. Reproduced with permission from Herndon (2018).

Thawing permafrost: human health hazard
In the Arctic, in remote areas of Siberia, outbreaks of anthrax have occurred. Carcasses of reindeer which were covered by permafrost have resurfaced causing a large outbreak of anthrax disease in 2016 in the Yamalo-Nenets region, in the Arctic Russian Siberia. The outbreak caused the death of one person and over 2000 reindeer (Stella *et al* 2020).

The state of the permafrost is intrinsically linked to the snow cover, with trends in permafrost temperature consistent with snow depth temporal variability (Stieglitz *et al* 2003). Snow is a good insulator—soil beneath a deep snowpack will be warmer than soil exposed to the air, and, depending on the timing of the formation of the snowpack, may contribute to warming of the permafrost (Park *et al* 2015, Ahmad and Painter 2020, Grünberg *et al* 2020). However, the high albedo of snow means that the majority of solar radiation is reflected, so conversely can prevent permafrost thaw (Yoshikawa *et al* 2020). Therefore it is imperative to understand the dynamics of the snow cover to understand changes in the permafrost and other surfaces covered by snow. The physical principles of snow and its evolution are described in the following chapter.

References

Anisimov O A *et al* 2007 *Climate Change 2007: Impacts, Adaptation and Vulnerability. Contribution of Working Group II to the Fourth Assessment Report of the Intergovernmental Panel on Climate Change* IPCC 653–85

Aye K-M *et al* 2019 Planet four: probing springtime winds on Mars by mapping the southern polar CO_2 jet deposits *Icarus* **319** 558–98

Barrett A M *et al* 2017 Clastic patterned ground in Lomonosov crater, Mars: examining fracture controlled formation mechanisms *Icarus* **295** 125–39

Barry R and Gan T Y 2011 *The Global Cryosphere: Past, Present and Future* (Cambridge: Cambridge University Press)

Dobinski W 2011 Permafrost *Earth-Sci. Rev.* **108** 158–69

French H M 1996 *The Periglacial Environment* 2nd edn (Harlow: Addison Wesley Longman)

Grünberg I *et al* 2020 Linking tundra vegetation, snow, soil temperature, and permafrost *Biogeosciences* **17** 4261–79

Herndon E M 2018 Permafrost slowly exhales methane *Nat. Clim. Change* **8** 273–4

Hjort J and Karjalainen O 2018 Degrading permafrost puts Arctic infrastructure at risk by mid-century *Nat. Commun.* **9** 5147

Hjort J and Streletskiy D 2022 Impacts of permafrost degradation on infrastructure *Nat. Rev. Earth Environ* **3** 24–38

Hope C and Schaefer K 2016 Economic impacts of carbon dioxide and methane released from thawing permafrost *Nat. Clim. Change* **6** 56–9

Hubbard B, Souness C and Brough S 2014 Glacier-like forms on Mars *Cryosphere* **8** 2047–61

Jakosky B M and Phillips R J 2001 Mars' volatile and climate history *Nature* **412** 237–44

Ahmad J and Painter S L 2020 2020 Permafrost thermal conditions are sensitive to shifts in snow timing *Environ. Res. Lett.* **15** 084026

Kessler M A and Werner B T 2003 Self-Organization of sorted patterned ground *Science* **299** 380–3

Key J, Drinkwater M and Ukita J 2007 *IGOS Cryosphere Theme Report* **1405** 100

Knoblauch C *et al* 2018 Methane production as key to the greenhouse gas budget of thawing permafrost *Nat. Clim. Change* **8** 309–12

Lantuit H and Pollard W H 2008 Fifty years of coastal erosion and retrogressive thaw slump activity on Herschel Island, southern Beaufort Sea, Yukon Territory, Canada *Geomorphology* **95** 84–102

Legendre M *et al* 2014 Thirty-thousand-year-old distant relative of giant icosahedral DNA viruses with a pandoravirus morphology *PNAS* **111** 4274–9

Ling F and Zhang T 2004 A numerical model for surface energy balance and thermal regime of the active layer and permafrost containing unfrozen water *Cold Reg. Sci. Technol.* **38** 1–15

Lucchitta B K 1981 Mars and Earth: comparison of cold-climate features *Icarus* **45** 264–303

Murton J B *et al* 2022 A multimethod dating study of ancient permafrost, Batagay megaslump, east Siberia *Quat. Res.* **105** 1–22

Nicolsky D J, Romanovsky V E and Panteleev G G 2009 Estimation of soil thermal properties using *in situ* temperature measurements in the active layer and permafrost *Cold Reg. Sci. Technol.* **55** 120–9

Paepe R, Hoover R and Van Overloop E 2001 Patterned ground as an evidence of water on Mars: A comparison with planet Earth *Permafrost Response on Economic Development, Environmental Security and Natural ResourcesNATO Science Series* ed R Paepe *et al* (Dordrecht: Springer) pp 581–8

Park H *et al* 2015 Effect of snow cover on pan-Arctic permafrost thermal regimes *Clim. Dyn.* **44** 2873–95

Peppin S S L and Style R W 2013 The physics of frost heave and ice-lens growth *Vadose Zone J.* **12** vzj2012.0049

Revich B, Tokarevich N and Parkinson A J 2012 Climate change and zoonotic infections in the Russian Arctic *Int. J. Circumpolar Health* **71** 18792

Schuur E A G and Abbott B 2011 High risk of permafrost thaw *Nature* **480** 32–3

Seppälä M 1982 An experimental study of the formation of palsas *Proc. 4th Can. Permafrost Conf.* pp 36–42

Stella E *et al* 2020 Permafrost dynamics and the risk of anthrax transmission: a modelling study *Sci. Rep.* **10** 16460

Stieglitz M *et al* 2003 The role of snow cover in the warming of Arctic permafrost *Geophys. Res. Lett.* **30** 1721

Taber S 1930 The mechanics of frost heaving *J. Geol.* **38** 303–17

Vaughan D G *et al* 2013 Climate change 2013: the physical science basis *Contribution of Working Group I to the Fifth Assessment Report of the Intergovernmental Panel on Climate Change. Observations: Cryosphere* pp 317–82

Yoshikawa K *et al* 2020 Current thermal state of permafrost in the southern Peruvian Andes and potential impact from El Niño–Southern Oscillation (ENSO) *Permafr. Periglac. Process.* **31** 598–609

IOP Publishing

Introduction to the Physics of the Cryosphere (Second Edition)

Melody Sandells and Daniela Flocco

Chapter 4

Snow

Snow is a fascinating material—it grows, evolves, densifies and melts, and there are many different crystal forms. Snow on the ground is very different from snow in the air, and is often spatially heterogeneous. The physics of snow extends far beyond what can be included in this chapter, so there is a lot more to find out! For example, these principles are used as the basis for avalanche research, or investigations into how forest cover affects snow evolution.

This chapter looks at why different snow crystal shapes form in the atmosphere, how a snowpack builds up and what happens to it over time. This is followed by how to use physics to determine the snowpack temperature, which governs how snow crystals change and whether the snow melts.

4.1 Snow crystal formation

Snow crystals are building blocks of the cryosphere. Grown in the atmosphere from dust or aerosol particle nuclei, the exact form of a snow crystal depends on the temperature and amount of moisture in the air, although the chemical composition of the atmosphere also has an effect (Libbrecht 2001). A range of crystals may form: the familiar depiction of a snowflake is the stellar dendrite, although other forms include hexagonal plates, dendrites with hexagonal plate extensions, needles, columns and capped columns. The crystals are hexagonal or exhibit sixfold symmetry in nature, from the shape of the H_2O molecular structure (see chapter 2, figure 2.2). Although most pictures of snowflakes show perfect symmetry, these are the supermodels of crystals. Most snow crystals are lopsided to some degree (Libbrecht 2005).

Figure 4.1 illustrates a Nakaya diagram, named after the researcher who first started attempting to grow snow crystals in the laboratory environment. Nakaya noticed that the type of crystal grown could be controlled with different temperature and humidity conditions. Supersaturation is the amount of moisture in the air above the maximum limit given by the Clausius–Clapeyron equation (2.6). With an

doi:10.1088/978-0-7503-3647-5ch4

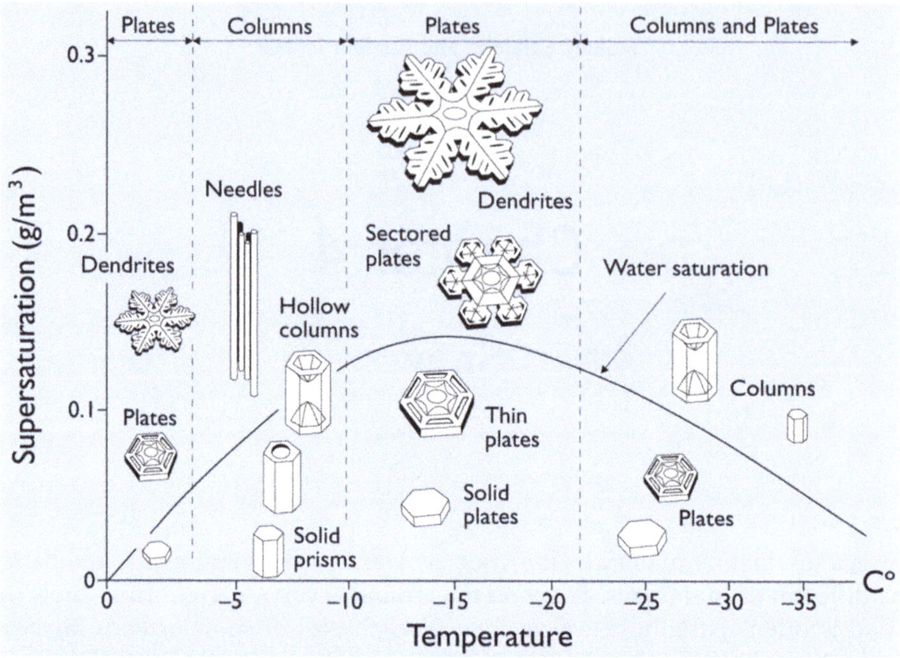

Figure 4.1. Precipitation snow crystal forms depend on temperature and moisture conditions. This is known as a Nakaya diagram. Reproduced by kind permission from Kenneth Libbrecht.

abundance of moisture, the vapour deposits onto lumps and bumps projecting from the crystal. The air is looking to shed the moisture, and shed it fast. Anything sticking out grows rapidly, leading to the formation of dendrites and columns. Such points are thermodynamically unstable, though, as will be discussed later.

Under moderate moisture conditions, and at a more leisurely pace, vapour tends to attach to dangling chemical bonds. This leads to the smoothing of surfaces and the subsequent formation of hexagonal plates. Electron microscope images of precipitation snow crystals are shown in figure 4.2. The change in crystal form with supersaturation is reasonably well understood. However, why totally different crystals form when the temperature is varied by a few degrees was not (Libbrecht 2005) until recently attributed to surface energy dynamics dependent on the width of facets (Libbrecht 2019).

4.2 Snowpack formation

Technically, a snowflake can be a single snow crystal or a collection of snow crystals stuck together. As snowflakes fall and deposit on the surface, they form a layer of snow. The properties of the snow layer change over time, depending on the thermal history of the snowpack, but dig down into snow and laid out before you is a snowfall record, as the layers are largely preserved. Figure 4.3 shows a backlit snow pit. The stratigraphy in the snow is evident.

Figure 4.2. Different forms of snow crystal found in snowfall. Electron microscope pictures by Electron and Confocal Microscopy Laboratory, Agricultural Research Service, US Department of Agriculture.

> **To dig a backlit snow pit**
> Dig two pits back-to-back, with a thin middle wall that is reasonably straight and smooth on either side. Allow the Sun to shine through the snow wall; you may want to cover the pits with a tarpaulin or whatever you have to hand to prevent sunlight shining directly down so that the layers show up more clearly. Alternatively, shine a light through the other side.

A simple way of modelling the snow would be to treat it as a single layer or a handful of fixed-width layers (this is the Eulerian grid system discussed in chapter 2). In practice, most snow models use a number of layers (some have hundreds) to represent the snow, and more sophisticated numerical solution methods are needed to simulate snowpack changes. Two main changes occur in the dry snow layers: densification and snow crystal growth.

4.3 Metamorphism and densification

Over time, the snowpack becomes more and more dense. For fresh snow, this is partly mechanical with breaking of dendritic[1] arms, but is also thermodynamical.

[1] A crystal that has branched in growth into two parts. Crystals that grow in this way (dendritic growth) have a branching tree-like appearance. *A Dictionary of Science*, 1999, Oxford University Press, Market House Books Ltd.

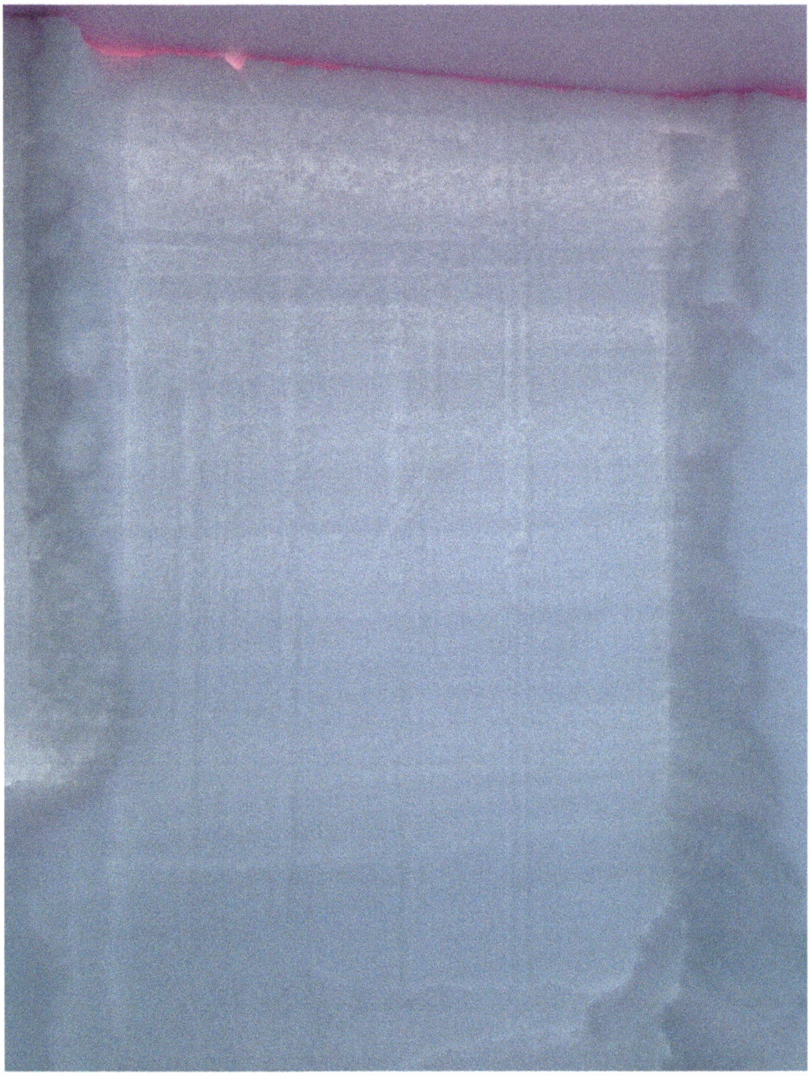

Figure 4.3. Backlit snow pit illustrating stratigraphic layers deposited in separate precipitation events.

Metamorphic changes in the snow crystal structure lead to better packing of snow grains and a settling of the snow. During this time, bonds are formed between snow grains and metamorphic densification slows in the more cohesive structure. As more snow accumulates, the pressure of the snow above forces closer packing of snow grains, but this rate of densification is slower than the initial rate. Some models, such as SNTHERM (Jordan 1991), model these two processes separately, through a change in snow layer thickness, Δz, for constant snow mass:

$$\left. \frac{1}{\Delta z}\frac{\partial \Delta z}{\partial t} \right|_{\text{metamorphism}} = -2.8 \times 10^{-6} \times e^{-0.04(273.15-T)}, \tag{4.1}$$

$$\left| \frac{1}{\Delta z} \frac{\partial \Delta z}{\partial t} \right|_{\text{overburden}} = -\frac{P_s}{\eta}. \tag{4.2}$$

Densification occurs much faster for wet snow, and the rate of densification by metamorphism in equation (4.1) is doubled. For higher-density snow ($\geqslant$150 kg m^{-3}), the densification process slows and the densification rate is then reduced by a factor, $e^{-0.046(\gamma_i - 150)}$, where γ_i is the mass of ice per unit volume of snow. Densification due to the weight of snow above is given by equation (4.2), where P_s is the pressure of overlying snow and η is the viscosity coefficient, which is dependent on the temperature and snow density (Kojima 1967). Jordan (1991) indicates that the viscosity also depends on grain type.

Other models may use a different approach, e.g. Snobal (Marks and Dozier 1992) uses a time-dependent curve to approximate the process:

$$\rho_s^t = \frac{\rho_{s,\,\max}^{\text{grav}}}{(1 + \tau t)}, \tag{4.3}$$

where $\rho_{s,\,\max}^{\text{grav}}$ is the maximum density under gravitational compaction only and τ is a constant that represents the time taken for half the maximum density to be reached (e.g. 10 days). A separate but similar equation is used to simulate more rapid densification in the presence of liquid water in Snobal.

Both approaches, as well as numerous alternative equations, are empirical to some degree and are based on limited observations. These equations describe densification up to a density of around 550 kg m^{-3}, where snow crystals are closely packed. Densification to ice (important for ice core analysis; see chapter 2) involves three stages of densification (Duval 1985, Herron and Chester 1980), of which close packing is the first. The second stage involves the growth of necks between particles, otherwise known as sintering (Colbeck 1989), and grain rearrangement (Maeno and Ebinuma 1983, Ebinuma and Maeno 1987). At the third stage, pores in the ice matrix are closed, and densification occurs through compression of the air bubbles (Herron and Chester 1980).

As noted by Jordan (1991), the snow microstructure has an effect on densification rates. Bonds formed between crystals affect the compressive strength of the snow (Hobbs 1965, Keeler 1969). Freitag *et al* (2004) and Gerland *et al* (1999) found densification rates that were dependent on snow microstructure. More recently, firn density changes measured by Morris and Wingham (2014) indicated that summer and winter snow has different strengths, and they proposed that densification rates depended on the temperature history of the snow. Densification is discussed further in the following chapter on land ice; see chapter 5.

Microcomputed tomography (e.g. Flin *et al* 2004, Heggli *et al* 2009 and Srivastava *et al* 2010) has allowed metamorphic changes in the snowpack structure to be investigated in the laboratory. These techniques allow reconstruction of the full 3D structure of snow for a relatively small sample size (e.g. 10 × 10 × 15 cm). However, none of these studies have used compressive loads comparable to those seen in natural snowpacks.

Snow densification is an important process, as snow ranges from light and fluffy to hard and solid. Densification should be included in a snow model. Mass is usually the value of interest, but depth is one of the easiest to measure in the field. As mass = density × volume, it is important to get the density correct. Snow depth also governs the temperature gradient and, in turn, the rate of snow crystal growth. The rate of heat transfer in the snow is parameterised by the thermal conductivity, a function of density, as described later in equation (4.9). Density also affects electromagnetic scattering of microwave radiation, the only method we have of measuring snow mass on a global scale (see chapter 1).

Wrong type of snow
Terry Worrall, *Evening Standard*, February 1991: 'We are having particular problems on this occasion with the type of snow, which is dry and powdery and is actually penetrating all the protection we had on some of our locomotives'. Unfortunately, UK rail companies are still ridiculed for blaming transport problems on 'the wrong sort of snow'. This illustrates how the density of snow impacts our lives.

One of the most fascinating areas of snow research is how snow crystals change. Not only do they start from different shapes, but they are also constantly evolving into other forms. Initial snow crystals can be small, e.g. around 0.2 mm in diameter, but over a season can grow as large as 6 mm or more. Figure 4.4 illustrates some

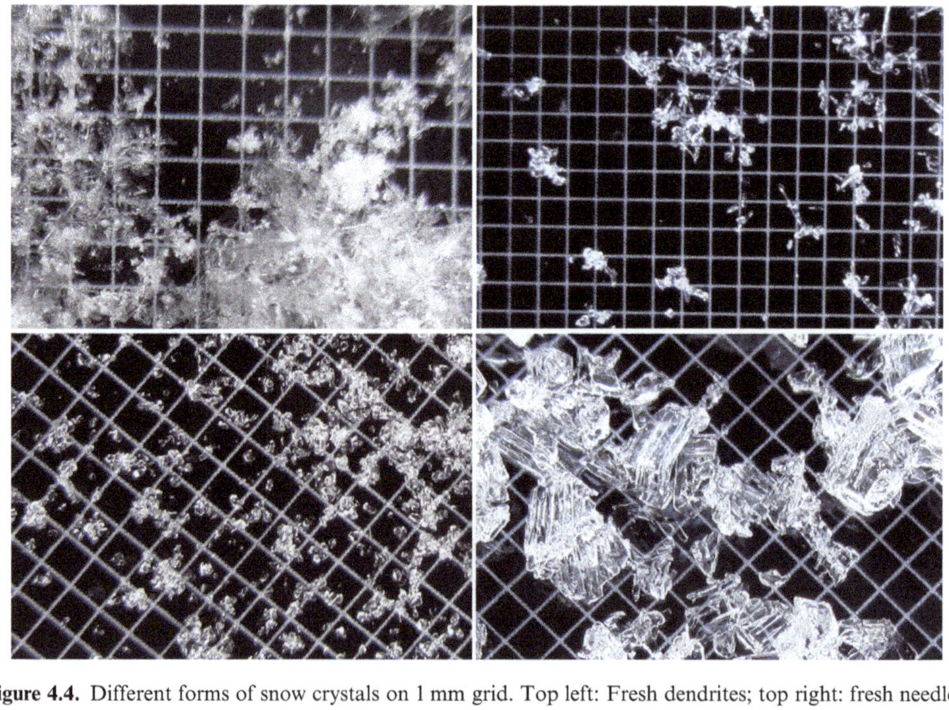

Figure 4.4. Different forms of snow crystals on 1 mm grid. Top left: Fresh dendrites; top right: fresh needles; bottom left: small rounded crystals; bottom right: large depth hoar.

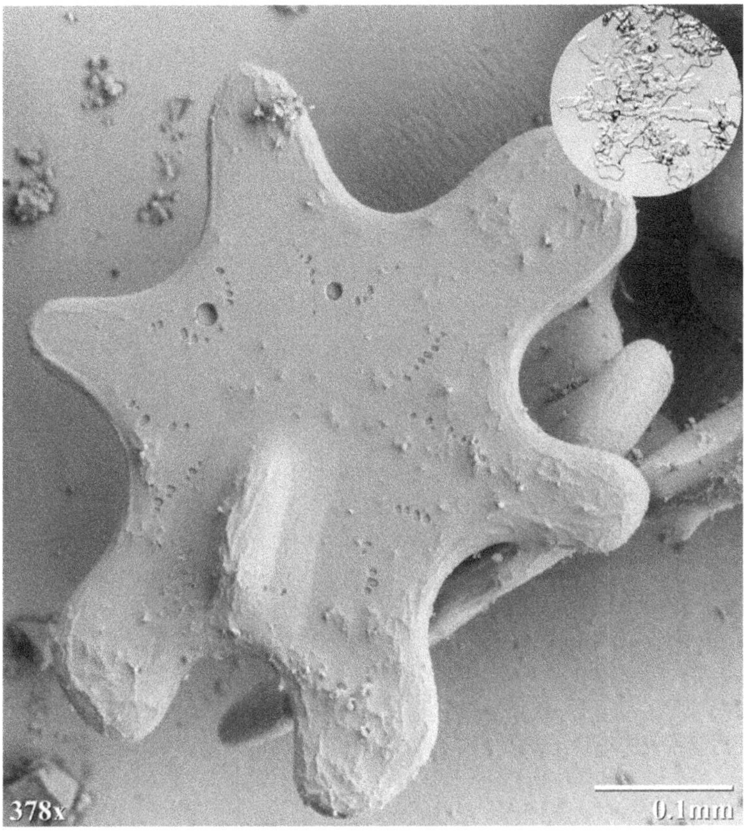

Figure 4.5. Rounded dendritic snow crystal imaged with a scanning electron microscope (main picture) and a similar crystal viewed through a light microscope (top right). Pictures by Electron and Confocal Microscopy Laboratory, Agricultural Research Service, U. S. Department of Agriculture.

different forms of snow found within a snowpack in January 2011 in Churchill, Canada.

The rate of crystal growth and the forms they take depend on the temperature gradient. Equi-temperature grain growth occurs under small or negligible temperature gradient conditions and acts to reduce the surface free energy (Sommerfeld and LaChapelle 1970), so snow grains become more rounded. As mentioned earlier, points and dendritic arms are thermodynamically unstable so lose their shape, as shown in figure 4.5. Colbeck (1998) suggested that equi-temperature metamorphosis should be renamed 'radius of curvature' metamorphism, although the term is still not in common use.

Radius of curvature metamorphism drives changes in snow grains because there is a pressure difference between phases for a curved surface. Laplace's equation[2] (Colbeck 1998) suggests that

[2] See also equation (3.1), which governs whether the frost freezing front will move down into soil or liquid water is drawn up to form ice lenses.

$$p_i = p_v = 2\frac{\sigma_{iv}}{r_{iv}}, \quad (4.4)$$

where p is the pressure for ice and vapour phases i and v, respectively, σ is the interfacial energy and r is the mean radius of the curvature of the interface (the mean of the radius in two orthogonal planes)[3]. Equation (4.4) illustrates that the pressure difference is larger for small radii of curvature, which leads to high rates of sublimation over points and edges. Vapour diffuses to areas of lower concentration, found over areas of larger radius of curvature, where condensation back into the ice phase occurs to maintain equation (4.4). This is a simplistic explanation as it ignores micro-gradients in temperature across the snow grain. Any changes in temperature result in larger rates of mass transfer, as the vapour pressure must also satisfy the Clausius–Clapeyron equation described in chapter 2, i.e. the amount of water that air can hold for a given temperature.

Rounded crystals similar to those in figure 4.6 are commonly found. Bonds (or 'necks') form between crystals, which have a role in the strength of the snow and densification. Under strong temperature gradients, however, a different type of grain growth occurs. Water sublimates from warmer regions and diffuses to colder areas, where it condenses. Matter moves from the top of one crystal, travels across the pore space and deposits on the face of another crystal. Yosida *et al* (1955) described this process as hand-to-hand transport. This process is dependent on the temperature gradient, the temperature and the density (Marbouty 1980). Crystals formed under these conditions are faceted, and can be poorly bonded in the case of depth hoar crystals formed at high-temperature gradients. Figure 4.7 illustrates the faceted nature of depth hoar crystals, although these can take many forms. A full description of all types of snow and ice is given in Fierz *et al* (2009).

Temperature gradient metamorphism has been observed with a computed microtomography machine, which gives a full 3D structure of the snowpack. This video demonstrates what happens to rounded snow crystals when a temperature gradient of 50 K m^{-1} is applied to snow for 3 weeks. However, the process is less well defined if the direction and magnitude of the temperature gradient is varied.

In nature, snow is always subject to temperature gradients for some period of time, unless the snowpack is 'ripe', where the temperature would be maintained at 0 °C throughout the snowpack during melting. Figure 4.8 illustrates a change in temperature gradient within one morning. These data were taken at slightly different locations, which is why there is some variation in snow depth. At this location, the ground was close to the freezing point, so the temperature gradient is largely negative at the base of the snowpack. Close to the surface, however, the sign and magnitude of the temperature gradient changes over the day. In order to simulate changes in the internal snow temperature, energy balance equations are used (and mass conservation equations, also, if there are any changes in phase).

[3] This equation does not take into account non-isotropic stresses.

Figure 4.6. Sintered rounded snow crystals. Pictures by Electron and Confocal Microscopy Laboratory, Agricultural Research Service, U. S. Department of Agriculture.

4.4 Mass and energy balance modelling

General energy conservation equations were introduced in chapter 2. In order to simulate the snow temperature profile for understanding snow grain size evolution, or for predicting subsurface melt, the snow must be considered as a series of layers, and conservation of energy calculated for each layer simultaneously. As described earlier, the snowpack is naturally layered with new layers deposited with each snowfall event. Most snow models have common physics principles behind them, but differ in their mathematical complexity. For example, SNTHERM (Jordan 1991) and SNOWPACK (Bartelt and Lehning 2002) create a new layer each time it snows. Other models require more computational efficiency, so have only two, three or five layers in a fixed Eulerian grid. With fewer layers, snow needs to be redistributed between the layers periodically, so care must be taken to transfer the energy also. Here, for simplicity we will assume that snow is preserved within the original layers, i.e. a Lagrangian grid (see chapter 2).

Within the snow, the temperature is controlled mainly by thermal conduction and phase changes. Some solar radiation penetrates below the snow surface, which can

Figure 4.7. An example of a depth hoar crystal formed under a strong temperature gradient imaged with a scanning electron microscope (main picture) and a similar crystal viewed through a light microscope (top left). Pictures by Electron and Confocal Microscopy Laboratory, Agricultural Research Service, U. S. Department of Agriculture.

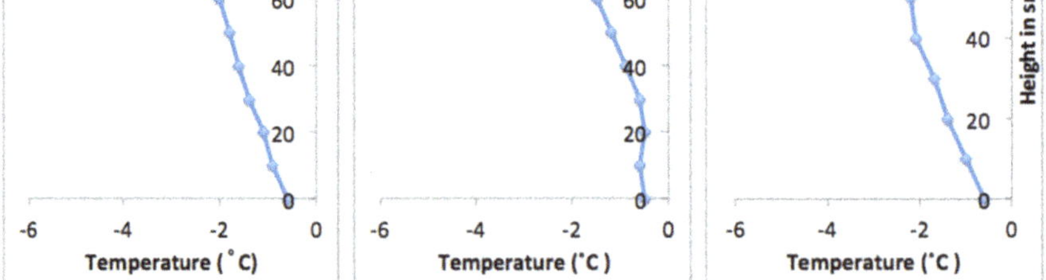

Figure 4.8. Variation in temperature profile in one morning. Data were taken at different locations within the Reynolds Mountain East catchment, Idaho, on 16 March 2001.

even lead to melt below the frozen surface (Koh and Jordan 1995). Some studies consider convective transport within the snow, also termed wind-pumping (Colbeck 1980, Clarke *et al* 1987, Albert and McGilvary 1992), but this is considered negligible here.

For an internal snow layer, the energy balance equation is given by equation (4.5) (Jordan 1991), where the first term, the change in enthalpy (heat stored), is given by the energy transported by mass flux (second term), the heat conduction (third term) and radiative energy flux (fourth term) across the boundary. For an internal layer, it is only solar radiation out of the full surface energy balance components in equation (2.10) that penetrates through the snow and contributes to the final term in equation (4.5):

$$\frac{\partial}{\partial t}\int_V \rho_t h_t dV = \overline{-\sum_k \int_V \mathbf{U}_k h_k \cdot d\mathbf{S}} + \overline{\int_S k_t \nabla T \cdot d\mathbf{S}} + \overline{\int_S \mathbf{I}_R \cdot d\mathbf{S}}. \tag{4.5}$$

Taking each term one at a time, starting with the term on the left-hand side (change in enthalpy with time), the density of snow is continually changing, e.g. as described in equations (4.1) and (4.2) or equation (4.3). Specific enthalpy, h, is the energy required to change the temperature of a unit mass of a substance from T to reference temperature T_0 under constant pressure. As this may involve a change in phase, the specific enthalpy may include a latent heat term. Relative to the freezing point of water (T_0 = 273.15 K), specific enthalpy is defined as

$$h = \int_{T_0}^{T} c(T)dT + L. \tag{4.6}$$

The specific heat capacity of ice, c, is temperature-dependent. An empirical representation of this is (Dickinson *et al* 1915)

$$c(T) = 2117.2 + 7.8(T - 273.15). \tag{4.7}$$

Although the second term in the energy balance equation (4.5) allows for mass flux of all phases, heat transport through the vapour phase may be considered negligible. Lagrangian grids that move with the compacting snow conserve the ice mass, so would have no ice mass flux across layer boundaries. The energy contribution of liquid water flux, $\mathbf{U}_l$, is an important contribution, and water flow is discussed in the following mass balance section.

Conduction of heat is given by the third term in equation (4.5) and is recognisable in its similarity to Fourier's law, which describes general heat conduction:

$$\frac{dq_z}{dt} = -k\frac{dT}{dz}. \tag{4.8}$$

In snow, the thermal conductivity depends on the structure of the snow, in particular the connections and bonds between the snow crystals and the size distribution of those crystals (Arons and Colbeck 1998, Löwe *et al* 2013). A common approach is to link the thermal conductivity of snow to the ice and air constituents, though the snow density. Sturm *et al* (1997) suggests the following equations for effective thermal conductivity (here density is given in g cm^{-3}):

$$k_{\text{eff}} = 0.023 + 0.234\rho_s \qquad \{\rho_s < 0.156\}$$
$$k_{\text{eff}} = 0.138 - 0.101\rho_s + 3.233\rho_s^2 \qquad \{0.156 \leqslant \rho_s \leqslant 0.6\}. \tag{4.9}$$

Effective thermal conductivity includes the effect of vapour transport within the snow layer. In practice, it is impossible to separate conduction through the solid phase and vapour flux effects experimentally to measure the snow thermal conductivity alone. Similar terms are lumped together to form the effective thermal conductivity of sea ice, as described in chapter 6. Domine *et al* (2011) suggested a modification of equation (4.9) to incorporate the shear strength, as would be predicted by avalanche models. In essence, the modification is related to the bond size between crystals.

Solar radiation is the only surface energy flux that penetrates to the snowpack interior. Chapter 2 introduced the concept of surface albedo, which is the proportion of solar radiation reflected by the surface. For snow, this is related to the snow grain size. Fresh snow appears bright because it has small crystal sizes and reflects the vast majority of sunlight. Actually, snow reflectance is dependent on the wavelength of light (as shown in figure 4.9), among other factors (Warren 1982). The data in figure 4.9 were taken just over one month apart, during which time the snow crystals grew and reflectance decreased. Remote sensing can exploit the change in reflectance at near-infrared wavelengths to determine snow grain size (e.g. Nolin & Dozier 2000; Picard *et al* 2009; Zege *et al* 2011).

A number of different methods may be employed to represent snow albedo (e.g. Marshall and Warren 1986, Kokhanovsky and Zege 2004, Melloh *et al* 2002). Essery *et al* (2013) gives an overview of all parameterisations of albedo, densification

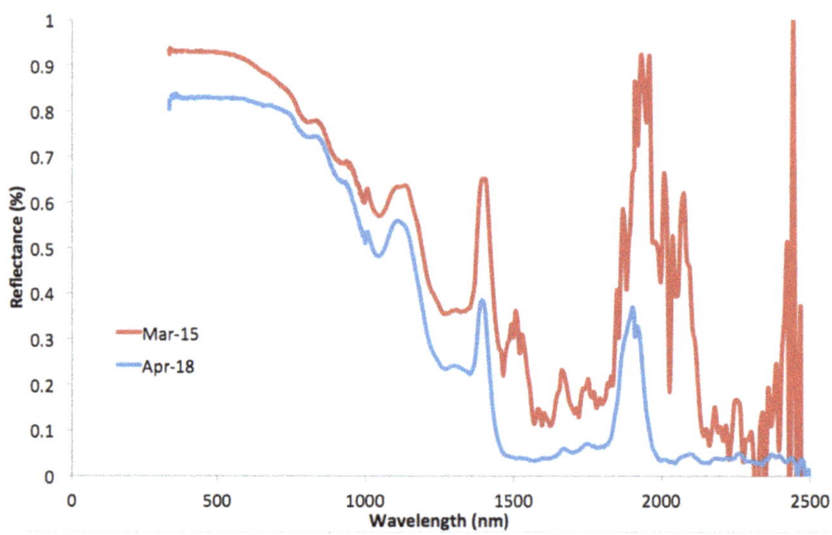

Figure 4.9. Sunlight reflected by the snow depends on the size of the snow crystals and the wavelength of light. Data are from the Reynolds Creek Experimental Watershed, Idaho, in 2001. Note that water vapour in the atmosphere absorbs radiation at around 1380, 1880 and 2680 nm, leading to noise in the signal.

and thermal conductivity used in snow models. Once the albedo is parameterised (modelled in some way or given an assumed value), the attenuation of solar radiation within snow layers can be determined from equation (4.10) (Jordan 1991). Different values for extinction coefficients are given for visible wavelengths (β_{vis}), which penetrate deeply in the snow, and near-infrared wavelengths (β_{nir}), which only penetrate the top 2 mm or so:

$$\begin{aligned} I_{\text{solar, net}}^{\text{top}} &= S^{\downarrow}(1 - \alpha)(1 - e^{-\beta_{vis}\Delta z^{top}} e^{-0.002\beta_{nir}}) \\ I_{\text{solar, net}}^{j+\frac{1}{2}} &= (1 - e^{-\beta_{vis}\Delta z^{j}}). \end{aligned} \tag{4.10}$$

The only parameter left that is required to solve the snow energy balance (equation (4.5)) and predict the temperature profile within the snow is the liquid water mass flux, U_l. This is determined as part of the mass balance.

As described in chapter 2, the snow mass balance can be applied to the snowpack as a whole or to its individual constituents: the ice, the liquid water and the water vapour. Provided that ice mass is not moved between model layers, the only change to ice mass within a layer is due to a change of phase. The mass lost from one phase is gained by another, e.g. the rate of melt, M_{li}, is opposite to the rate of freezing, M_{il}:

$$M_{li} = -M_{il}. \tag{4.11}$$

Equation (4.12) shows the conservation equation for liquid water for an interior snow layer (Jordan 1991). The change in liquid mass equals the mass influx through the top of the layer, minus the mass draining from the bottom of the layer, plus the amount of snow that has melted, minus the amount of liquid water that has evaporated:

$$\frac{\partial}{\partial t}(\gamma_l \Delta z) = U^{\text{upper}} - U^{\text{lower}} + M_{li}\Delta z - M_{vl}\Delta z. \tag{4.12}$$

Liquid water flow through the snow is an interesting physics problem, which was described in detail by Morris (1987). One approach is to assume that drainage is only due to gravity, as did Jordan (1991). In that case, liquid water flux is given by equation (4.13):

$$U_l = \gamma_l v_l = -\frac{K_l}{\mu_l}\rho_l^2 g, \tag{4.13}$$

where μ_l is the dynamic viscosity (a constant) and K_l is the hydraulic permeability (ease of liquid water flow), given by equation (4.14):

$$K_l = 0.077 d^2 e^{-0.0078\gamma_i} s_e^3. \tag{4.14}$$

Note that the hydraulic permeability is dependent on the effective saturation s_e, the ice matrix density, but also the snow grain size. Yamaguchi *et al* (2010) confirmed the dependence of water flow on snow grain size. However, water flow often occurs through preferential drainage channels, or can be impeded at the boundary between layers, as discussed in detail by Marsh (1991). The laboratory experiments of Waldner *et al* (2004) distinguished matrix flow versus preferential flow as a function of the snow texture. Flow channels are instabilities that, once initiated, grow both deeper and wider. Drainage channels can be seen through dye experiments, as shown in figure 4.10.

Snow evolves over the season. It builds up through snowfall events, the layers compact and the individual crystals change shape and size. As the hours of daylight extend and the Sun gets higher in the sky, more energy is input into the snow. Spring brings complete melt of the seasonal snow cover. The physically based models of

Figure 4.10. Dye tracer experiment. Photo reproduced with kind permission from Jennifer Petrzelka. Copyright © 2010 John Wiley & Sons, Ltd, Jennifer Petrzelka.

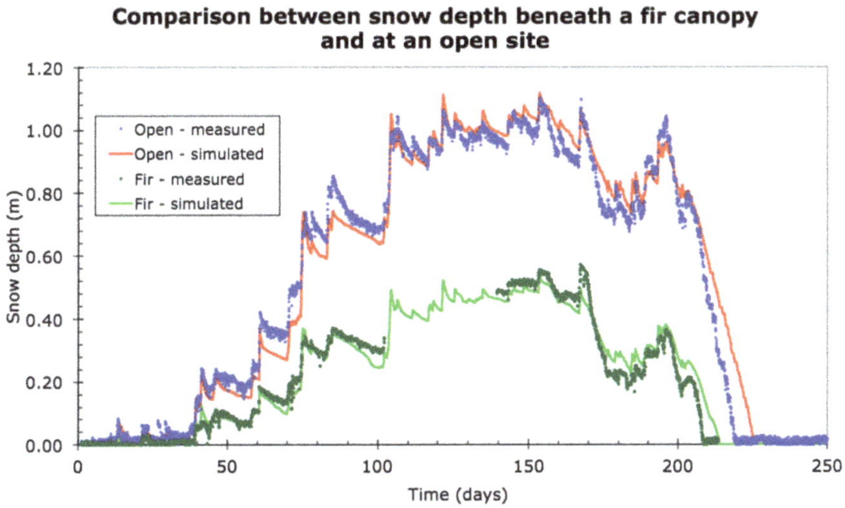

Figure 4.11. Simulation of snow depth at an open site and underneath a fir canopy at the Reynolds Creek Experimental Watershed, Idaho, 2001.

snow that we have do a reasonable job of simulating changes in the snow cover, as illustrated by Tribbeck *et al* (2006) and figure 4.11.

At high latitudes or elevations, though, the energy input may not be enough to melt the snow, and it may persist throughout the year. Snow surviving a melt season evolves into an intermediate stage between snow and glacial ice (described in the following chapter) known as *firn*. Firn continues to evolve due to the same physical processes, and at sufficiently high density becomes land ice. The next chapter discusses the physics of land ice.

References

Albert M R and McGilvary W R 1992 Thermal effects due to air flow and vapor transport in dry snow *J. Glaciol* **38** 273–81

Arons E M and Colbeck S C 1998 Effective medium approximation for the conductivity of sensible heat in dry snow *Int. J. Heat Mass Transf* **41** 2653–66

Bartelt P and Lehning M 2002 A physical SNOWPACK model for the Swiss avalanche warning: part I: numerical model *Cold Reg. Sci. Technol.* **35** 123–45

Clarke G K C, Fisher D A and Waddington E D 1987 Wind pumping: a potentially significant heat source in ice sheets *The Physical Basis of Ice Sheet Modelling* vol 70 (Wallingford: IAHS Publications) pp 169–80

Colbeck S C 1998 Sintering in a dry snow cover *J. Appl. Phys.* **84** 4585–9

Colbeck S C 1980 Thermodynamics of snow metamorphism due to variations in curvature *J. Glaciol.* **26** 291–301

Colbeck S C 1989 Air movement in snow due to windpumping *J. Glaciol* **35** 209–13

Dickinson H C and Osborne N S 1915 Specific heat and heat of fusion ofice *Bulletin of the Bureau of Standards* **248**

Domine F *et al* 2011 *J. Geophys. Res. Earth Surf.* **116** D00R06

Duval P 1985 Grain growth and mechanical behaviour of polar ice *Ann. Glaciol* **6** 79–82
Ebinuma T and Maeno N 1987 Particle rearrangement and dislocation creep in a snow-densification process *J. Phys. Colloq* **48** C1-263–1-269
Essery R *et al* 2013 A comparison of 1701 snow models using observations from an alpine site *Adv. Water Resour.* **55** 131–48
Fierz C *et al* 2009 The international classification for seasonal snow on the ground *Technical Report* IHP-VII Technical Documents in Hydrology No. 83/IACS Contribution No. 1 UNESCO/IHP
Flin F *et al* 2004 Three-dimensional geometric measurements of snow microstructural evolution under isothermal conditions *Ann. Glaciol.* **38** 39–44
Freitag J, Wilhelms F and Kipfstuhl S 2004 Microstructure-dependent densification of polar firn derived from X-ray microtomography *J. Glaciol.* **50** 243–50
Gerland S *et al* 1999 Density log of a 181 m long ice core from Berkner Island, Antarctica *Ann. Glaciol.* **29** 215–9
Heggli M, Frei E and Schneebeli M 2009 Snow replica method for three-dimensional X-ray microtomographic imaging *J. Glaciol* **55** 631–9
Herron M M and Langway C C Jr 1980 Firn densification: an empirical model *J. Glaciol* **25** 373–85
Hobbs P V 1965 The effect of time on the physical properties of deposited snow *J. Geophys. Res.* **70** 3903–7
Jordan R 1991 A one-dimensional temperature model for a snow cover: technical documentation for SNTHERM.89 *Technical Report* 91-16 U.S. Army Corps of Engineers: Cold Regions Research & Engineering Laboratory
Keeler C M 1969 The growth of bonds and the increase of mechanical strength in a dry seasonal snow-pack *J. Glaciol.* **8** 441–50
Koh G and Jordan R 1995 Sub-surface melting in a seasonal snow cover *J. Glaciol.* **41** 474–82
Kojima K 1967 Densification of seasonal snow cover *Proc. Physics of Snow and Ice* **1** 929–52
Kokhanovsky A A and Zege E P 2004 Scattering optics of snow *Appl. Opt.* **43** 1589–602
Libbrecht K G 2001 Morphogenesis on ice: the physics of snow crystals *Eng. Sci.* **64** 10–9
Libbrecht K G 2005 The physics of snow crystals *Rep. Prog. Phys.* **68** 855
Libbrecht K G 2019 A quantitative physical model of the snow crystal morphology diagram http://arxiv.org/abs/1910.09067 [cond-mat, physics:physics]
Löwe H, Riche F and Schneebeli M 2013 A general treatment of snow microstructure exemplified by an improved relation for thermal conductivity *Cryosphere* **7** 1473–80
Maeno N and Ebinuma T 1983 Pressure sintering of ice and its implication to the densification of snow at polar glaciers and ice sheets *J. Phys. Chem.* **87** 4103–10
Marbouty D 1980 An experimental study of temperature gradient metamorphism *J. Glaciol.* **26** 303–12
Marks D and Dozier J 1992 Climate and energy exchange at the snow surface in the Alpine Region of the Sierra Nevada: 2. Snow cover energy balance *Water Resour. Res.* **28** 3043–54
Marsh P 1991 Water flux in melting snow covers *Adv. Porous Media* **1** 61–124
Marshall S and Warren S G 1986 Parameterization of snow albedo for climate models *Proc. Snow Watch '85 (College Park, MD, 28–30 October)*
Melloh R A *et al* 2002 An efficient snow albedo model for the open and sub-canopy *Hydrol. Process.* **16** 3571–84
Morris E M and Wingham D J 2014 Densification of polar snow: measurements, modeling, and implications for altimetry *J. Geophys. Res. Earth Surf* **119** 349–65

Morris E M 1987 Modelling of water flow through snowpacks *Seasonal Snowcovers: Physics, Chemistry, Hydrology (NATO ASI Series)* eds H G Jones and W J Orville-Thomas vol 211 (Dordrecht: Springer) pp 179–208

Nolin A W and Dozier J 2000 A hyperspectral method for remotely sensing the grain size of snow *Remote Sensing of Environment* **74** 207–16

Picard G, Arnaud L, Nolin A W, Domine F, Nolin A W, Fily M and Nolin A W 2009 Determining snow specific surface area from near-infrared reflectance measurements: numerical study of the influence of grain shape *Cold Regions Science and Technology* **56** 10–7

Sommerfeld R A and LaChapelle E R 1970 The classification of snow metamorphism *J. Glaciol* **9** 3–18

Srivastava P K *et al* 2010 Observation of temperature gradient metamorphism in snow by X-ray computed microtomography: measurement of microstructure parameters and simulation of linear elastic properties *Ann. Glaciol.* **51** 73–82

Sturm M *et al* 1997 The thermal conductivity of seasonal snow *J. Glaciol* **43** 26–41

Tribbeck M J, Gurney R J and Morris E M 2006 The radiative effect of a fir canopy on a snowpack *J. Hydrometeorol.* **7** 880–95

Waldner P A *et al* 2004 Effect of snow structure on water flow and solute transport *Hydrol. Process.* **18** 1271–90

Warren S G 1982 Optical properties of snow *Rev. Geophys.* **20** 67–89

Yamaguchi S *et al* 2010 Water retention curve of snow with different grain sizes *Cold Reg. Sci. Technol.* **64** 87–93

Yosida Z *et al* 1955 Physical studies on deposited snow. Thermal properties *Contrib. Inst. Low Temp. Sci.* **7** 19–74

Zege E P *et al* 2010 Algorithm for retrieval of the effective snow grain size and pollution amount from satellite measurements *Remote Sensing of Environment* **115** 2674–85

Melody Sandells and Daniela Flocco

Chapter 5

Land ice

5.1 Definition and importance

The Earth's surface is covered by six million square kilometres of ice, corresponding to about 10% of the total surface. More than 99% of the volume of freshwater ice is stored in ice sheets (of which 90% is in the Antarctic) and the remaining in glaciers. Despite the relatively small area covered by ice, the processes that involve land ice and ice sheets have a strong impact on a number of environmental and human issues. Regions covered in ice are generally classified depending on their extent, as described in chapter 2. In this chapter, we discuss the physics of glaciers and ice sheets. Ice sheets (i.e. Greenland and Antarctica) are masses of ice in the polar regions and cover an area greater than 50 000 km^2. They can be as thick as several kilometres. Glaciers only form on land, can be rather small and are constrained by the underlying and surrounding topography. Many glaciers were formed during the last glaciation and can be found in the polar regions, or at low latitudes only on high mountains where they form above the snow line[1].

5.2 Glaciers

Around 20 000 years ago, during the last glacial period (Würm glaciation—115 000 to 11 700 years ago) land ice covered around 30% of the continents, so that most glaciers present today on Earth are residuals of the last glaciation (figure 5.1). Glaciers can be broadly classified depending on their location (e.g. mountain, valley, tidewater glaciers). Mountain glaciers form in high-altitude valleys or in depressions called cirques. Glaciers that terminate at the sea surface are classified as tidewater glaciers.

A number of spectacular glaciers are located in Chile. The Los Glaciares National Park is an archipelago in the southernmost part of Chile and contains almost 50 glaciers (e.g. figure 5.2, top). These are receding on average (Glasser *et al* 2011),

[1] The snow line is the altitude over which snow can survive over the summer.

doi:10.1088/978-0-7503-3647-5ch5

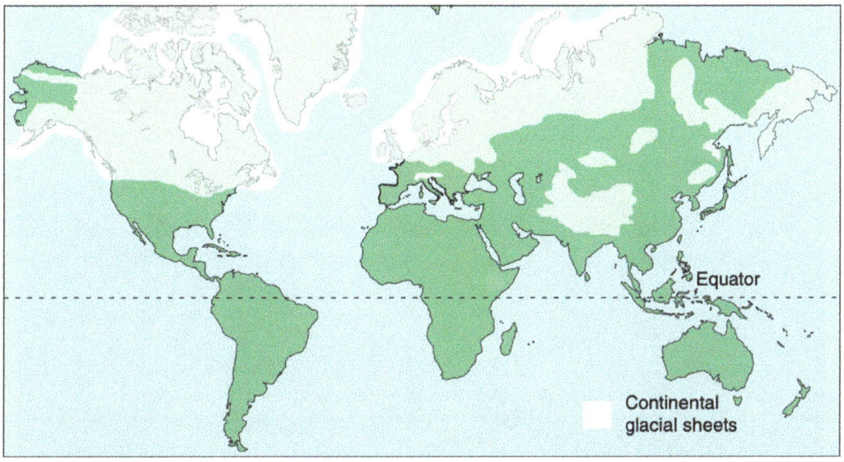

Figure 5.1. Extent of major glaciers at the height of the last ice age (20 000 BP) (27% of the Earth's land surfaces were covered by ice at that time). Reprinted from Sandells and Flocco (2014), Copyright © 2014 Morgan & Claypool Publishers.

but a few of the glaciers in the park are advancing. Small glaciers are concentrated in the northern part of the park, whilst larger ones such as the Perito Moreno (figure 5.3) are located in the southern part. The Perito Moreno is the third largest reservoir of freshwater on Earth and is remarkable. The glacier flows into Lake Argentino and is always in movement: close to the lake it is famous for its crumbling edge creating spectacular and dangerous scenery, but at the top it grows, thereby keeping its size stable.

5.2.1 Morphological aspects

Glaciers can be divided in two different zones: an accumulation zone and an ablation or wastage zone, as illustrated in figure 5.4. These zones are separated by an equilibrium line. Net snow accumulation occurs above the snow line (which depends on latitude and elevation) and leads to the growth or maintenance of a glacier. The area where the glacier originates is defined as its *head* and its lower part as its *terminus*.

The presence of a glacier heavily impacts the geography of an area, and even if the glacier disappears the signs of its previous existence are easily recognisable. When a glacier retreats, the morphology of the now-exposed terrain has some common structures such as glacial cirques, moraines, drumlins, truncated spurs and glacial erratics.

Glacial cirques. The formation of cirques is due mainly to erosion processes such as quarrying or plucking by the glacier. They are depressions of circular or elliptic shape with sills on the downhill, open side of the depression (figure 5.5). Small lakes can form in the cirques, as shown in figure 5.6.

Moraines. These can be found on the side of the glaciers (termed lateral moraines; see figure 5.7) or at the base (end moraines). They are dark and often look like small mounds or ridges. Moraines are formed by debris that is left behind once the glacier

Figure 5.2. (a) Glacier Italia, February 2014. Photograph courtesy of Sabrina Cortese. (b) Hubbard Glacier (Alaska), 10 December 2008, courtesy of Michele Barone.

disappears or is entrained within in it. Lateral moraines are parallel to the glacier's flow, while end moraines follow the shape of the glacier's terminus. They can also give information on whether the glacier is stable (terminal moraines) or receding (receding moraines). In the latter case they can have a typical amphitheatre shape.

Drumlins. Drumlins are present in groups and have an upturned spoon shape (figure 5.8). They form when a glacier over tills or moraines exerts a strong shear force and therefore are parallel to the flow direction. Exposed drumlins can be used

Figure 5.3. Perito Moreno. Reproduced, with kind permission, from latindiscoveries.com. Copyright © Photographer MIKEMDQ, Attribution-ShareAlike 4.0 International (CC BY-SA 4.0).

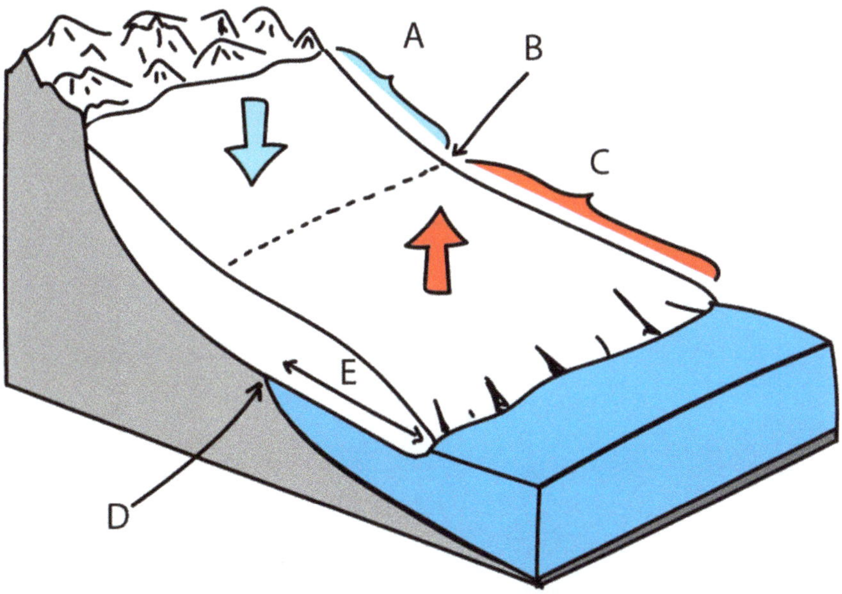

Figure 5.4. Features of a marine-terminating glacier. (A) Accumulation zone, (B) equilibrium line, (C) ablation zone, (D) grounding line (last point where glacier is in contact with the ground) and (E) ice shelf.

to deduce the path of the ice once the ice has gone. Most drumlins were formed during the last glacial era.

Truncated spurs. A glacier often has tributary glaciers feeding into it. If the glacial system disappears, the cross-section of the tributary glacier(s) can leave a V-shaped valley from which a waterfall can often be seen emerging, as shown in figure 5.9. The V-shaped base is an indication that glacier erosion has occurred.

Figure 5.5. Lower Curtis Glacier in the North Cascades in the State of Washington. The 1985 terminus position of this cirque glacier is shown in red. Copyright © Mauri S. Pelto, Ph.D.

Figure 5.6. Easedale Tarn, Lake District, UK—a lake formed in the depression left by a glacial cirque. Glacial erratics and steep slopes due to glacier plucking are also evident in this photo. Reproduced, with kind permission, from Ixia Sandells.

Figure 5.7. Lateral moraine on a glacier joining the Gorner Glacier, Zermatt, Switzerland. The Gorner Glacier runs along the bottom of the picture. The moraine bank runs up the left-hand side of the picture and results from rocks and Earth falling onto the glacier and from rocks being pulled out by the moving ice. If the glacier then melts a little the moraine bank is left clearly visible. Photograph taken by Adrian Pingstone in 1963. Copyright © Photographer/Uploader Adrian Pingstone, Public Domain.

Figure 5.8. Drumlins around Horicon Marsh in Wisconsin. Copyright © Photographer Doc Searls, Attribution 2.0 Generic (CC BY 2.0).

Figure 5.9. The Bridalveil Fall in Yosemite National Park, falling from a truncated spur. Photograph taken by Daniel Mayer in 2002. Released under the Creative Commons Attribution-ShareAlike 3.0 Unported licence.

Glacial erratics. These are rocks that the glacier has carried away from their original location and deposited elsewhere. This is impressive as the rocks can weigh tonnes. Their composition is important because knowledge of their original location can be used to trace back the original flow of the glacier.

5.2.2 Thermodynamic and dynamic aspects of a glacier

The constituents of glacial ice are mainly crystals or a matrix of ice, but also air voids and detritus. When snow starts to accumulate on the ground it is very fluffy and can have a density as low as 40 kg m^{-3} due to the high volume of air between the crystals. However, snow starts compacting very quickly under its own weight. As described in chapter 4, initial snow densification occurs because of mechanical breaking of dendritic arms, the rearrangement of snow grains under pressure into a more stable close-packed structure and from metamorphic changes to the crystal structure. This forms a more compact snow mass, called granular snow, with a density of about 400 kg m^{-3}. The expulsion of air and densification continues until the compacted snow is considered to be ice. A very long accumulation/compaction time and a pressure equivalent to some tens of metres of snow are required to produce glacial ice with a density of ~900 kg m^{-3}.

Firn can be considered an intermediate stage between snow and ice and covers almost the entire surface of ice sheets (Ligtenberg *et al* 2011). Firn densification is still a poorly understood process and most models are semiempirical. For example, models may use a temperature diffusion equation to give the snow temperature at depth and an empirical equation to determine densification rates that are dependent on the temperature. Here we show the expression proposed by Barnola *et al* (1991), which has been shown to be in good agreement with Antarctic observations (Ligtenberg *et al* 2011):

$$\frac{\partial T}{\partial t} = k\frac{\partial^2 T}{\partial z^2}, \tag{5.1}$$

$$\frac{d\rho}{dt} = \begin{cases} 11\Delta z\rho(\rho_i - \rho)e^{-\frac{E}{RT_s}}, & \text{for } \rho < 550 \text{ kg m}^{-3} \\ A_0\Delta z\rho\frac{t_{yr}}{b}f_e(\rho)(\Delta P)^3 e^{-\frac{E}{RT_s}}, & \text{for } \rho > 550 \text{ kg m}^{-3}. \end{cases} \tag{5.2}$$

Equation (5.1) is a simplified temperature diffusion equation that ignores heat advection and internal heating due to ice friction. In equation (5.1) k is the coefficient of thermal conductivity. In equation (5.2) E and A_0 are constants, Δz is the thickness of the firn layer, f_e is a mean analytical function related to the density, $t_{\rm yr}$ is the number of seconds in a year and P the pressure at each height due to the overlaying firn. This model considers processes related to percolation as well. A more general approach that models dry snow compaction, including the relationship with strain rate in the snow and overburden weight, was proposed by Morris and Wingham (2014).

It is important to underline that over a certain depth the pressure due to the ice weight determines the temperature at which melt occurs[2], as described in chapter 2. It is this rather than the surface energy balance that governs the thermodynamic behaviour of the ice interior at depth, and meltwater may exist at the base of glaciers that are below 0 °C. Nevertheless, glaciers can be classified according to their thermal regime and can be defined as *cold, warm* or *polythermal.* Cold glaciers are characterised by temperatures at the base that are below the pressure melting point during the year over the whole glacier's extent. Surface melt and runoff may occur

[2] The pressure melting point is the temperature at which ice melts at a given pressure.

on a cold glacier. In contrast, warm glaciers have a basal temperature that is high enough to cause melt at their bottom. When a glacier has separate regions with temperatures above and below the pressure melting point, it is called polythermal.

The thermal regime of a glacier governs its future: the presence of meltwater at the glacier's base speeds up its movement by lubricating the interface between ice and bedrock. These fast flows are called ice surges and can be as fast as 100 times the normal glacier speed. Ice surges can either be an isolated event or cyclic for some glaciers. Not only can basal water be present because of high temperatures or pressure melting (perhaps enhanced by friction), but also geothermal flux under the bedrock may cause the bottom of the glacier to melt. Topographic effects may increase the local geothermal heat flux by as much as 100% (Veen *et al* 2007).

A glacier generally behaves like a plastic[3] body (at thicknesses greater than 50 m) and deforms under shear stress until it starts flowing. Glacial flow is governed by the Glen–Nye flow law, which is depicted in figure 5.10:

$$F = k\tau^n, \tag{5.3}$$

where F is the flow rate (strain), τ is the stress and k and n are constants depending on temperature.

At the bottom and sides of the glacier, friction acts to reduce flow. Differential flow rates within the glacier lead to local deformation of the glacier. The surface velocity of a glacier, v_{surf}, is given by the Somigliana (1921) formula:

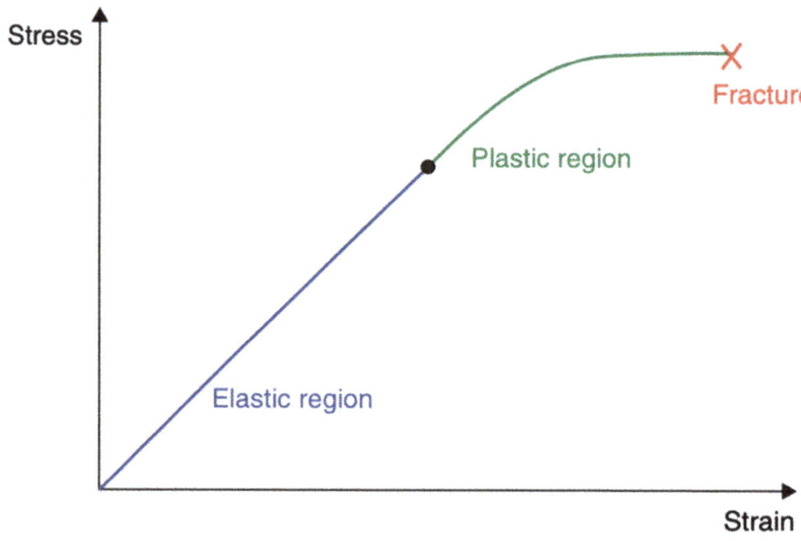

Figure 5.10. The stress–strain relationship of plastic flow (green section): a small increase in stress creates an exponentially greater increase in strain, which equates to deformation speed. Copyright © Author Moondoggy, Uploader by Pbroks13, Attribution-ShareAlike 3.0 Unported (CC BY-SA 3.0).

[3] In physics and materials science, plasticity describes the deformation of a material undergoing nonreversible changes of shape in response to applied forces (Lubliner 2008).

$$v_{\text{surf}} = \frac{\rho g \sin \alpha L^2 M^2}{2\mu(L^2 + M^2)}, \tag{5.4}$$

where ρ represents the ice density, μ the ice viscosity and α the bedrock slope, while L and M are the measures of the half-ellipsoidal cross-section that characterises the glacier. Equation (5.4) has been applied in the past to evaluate glacier depth but today remote-sensing methods are used.

The basal sliding of the glacier is described by the following equation:

$$\tau_D = \rho g h \sin \beta, \tag{5.5}$$

where τ_D is the shear rate, β is the glacier's surface slope in radians and g and h are the gravitational acceleration and the ice thickness, respectively. However, a glacier is not a perfect plastic body and under low levels of stress can sometimes break, creating cracks and crevasses (following brittle body behaviour). The location and direction of the crevasses indicate the underlying bedrock characteristics.

5.3 Where land and ocean meet: ice sheets and ice shelves

As described previously, ice sheets sit on land and extend to the sea where the ice floats, forming ice shelves. The mass balance of ice sheets is driven by the interaction of dynamic and thermodynamic processes (figure 5.11).

The evolution of the surface height H of the ice sheet is described by the following equation:

$$\frac{dH}{dt} = v_{\text{acc}} + v_{\text{me}} + v_{\text{fc}} + v_{\text{ice}} = v_{\text{b}}, \tag{5.6}$$

where v_{acc}, v_{me}, v_{fc}, v_{ice} and v_{b} are the vertical components of velocity that represent accumulation, melt, firn compaction, ice flow and bedrock movement, respectively

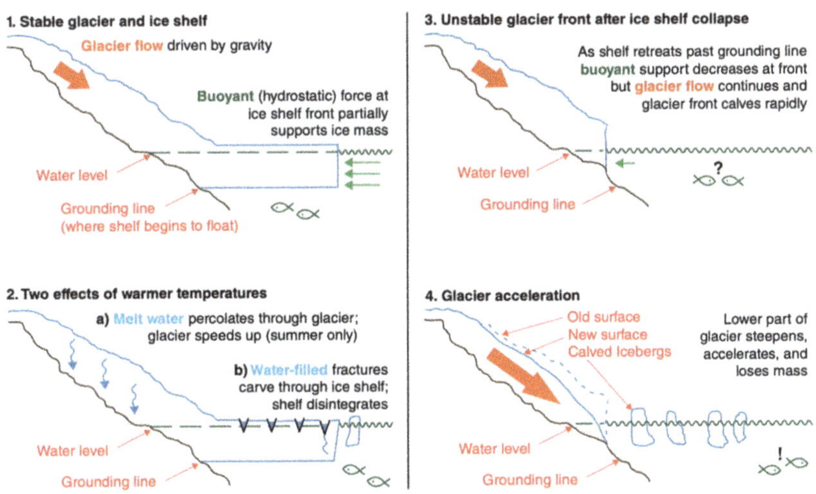

Figure 5.11. Factors controlling the flow of ice. Image courtesy of NASA.

(Ligtenberg *et al* 2011). The change of the surface topography of an ice body (e.g. snow accumulation) affects its gravitational flow.

Glaciers, constrained by topography, move along preferential streamlines towards the coastline. These are called ice streams and flow at a higher speed relative to the rest of the ice sheet. Reductions in ice sheet extent occur mainly as a result of increased flow when icebergs calve, but warmer air and ocean temperatures enhance both bottom and surface melting rates. Figure 5.12 shows the ice stream field in Antarctica retrieved by satellite observations (Bamber *et al* 2000).

One of the main features to look at for the assessment of the 'health' of a tidewater glacier is the behaviour of the grounding line. This is the place where the glacier base is no longer connected to the bedrock and the ice starts floating due to buoyancy and forms an ice shelf. Thicker ice is heavier, therefore the grounding line occurs closer to the ocean. If the accumulation rate over a glacier decreases the grounding line will retreat and the ice shelf will have a larger floating region with enhanced probability of iceberg calving. Iceberg calving does not have a direct impact on seawater levels, but an indirect effect is caused by the disappearance of the buttressing effect that the floating tongues have on the glaciers feeding the ice shelves. Glaciers flowing towards the coast tend to accelerate their course when no ice shelf is present to slow them down.

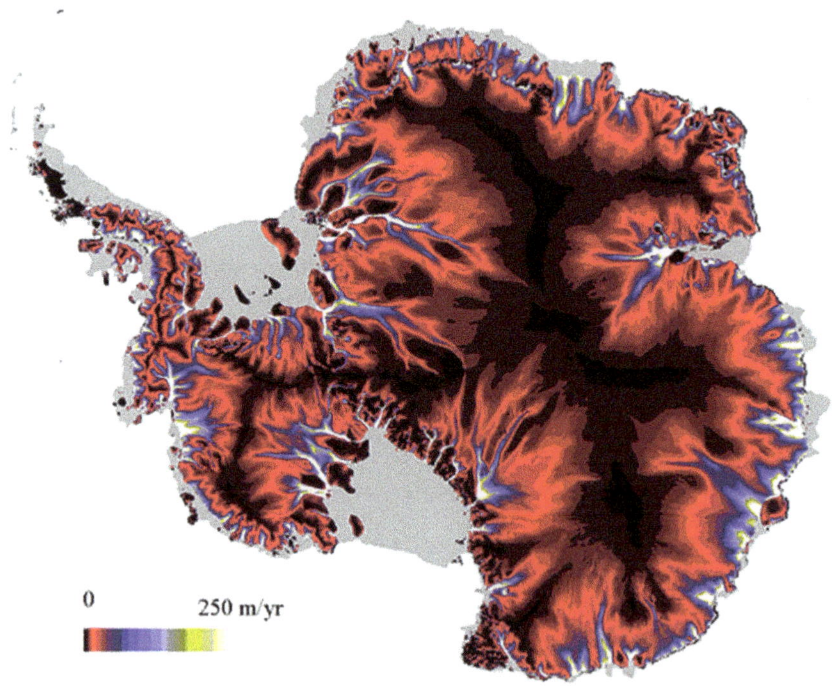

Figure 5.12. The ice stream field in Antarctica retrieved by satellite observations. Reproduced by permission of Professor Jonathan Bamber, Bristol University, UK. Copyright © Polargeo at English Wikipedia & Jonathan Bamber, Attribution-ShareAlike 3.0 Unported (CC BY-SA 3.0).

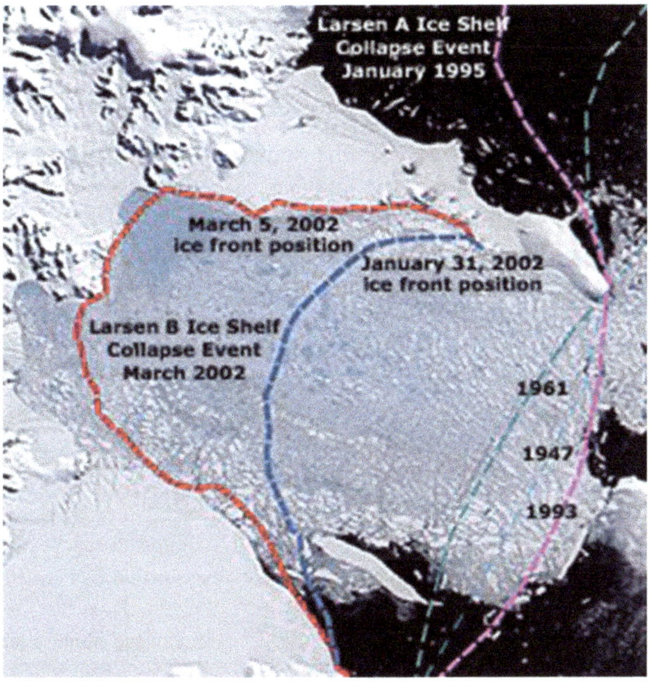

Figure 5.13. Coloured lines mark the Larsen B Ice Shelf edge in 1947, 1961, 1993 and 2002. NASA MODIS image courtesy of Ted Scambos, National Snow and Ice Data Center, University of Colorado, Boulder.

If the West Antarctic ice sheet were to melt completely this would cause a global sea level rise of 4.8 m (Bamber and Riccardo 2009).

An example of ice-shelf disintegration took place in 2002 in the area of the Larsen Ice Shelf, where one of the largest ice-shelf segments ever observed disintegrated in a very short time (figure 5.13). Prior to the calving, a large number of melt ponds were observed on the surface of the ice shelf. Their presence had enhanced the surface heat flux, but detachment of the 200 m thick iceberg was actually triggered by the sudden drainage of the ponds. Banwell *et al* (2013) shows how the water drained at the bottom of the ice sheet lifted the ice from the bedrock, initiating a chain reaction that caused the over 2000 melt lakes present on the surface to drain suddenly onto the bedrock, leading to the breakaway of the Larsen B ice shelf. The history of the Larsen B ice-shelf breakup is given by the National Snow and Ice Data Centre and can also be watched as a sequence.

Bottom melt of an ice shelf has further repercussions for the interface between the ice and the ocean: when ice shelves melt they create a layer of colder and fresher water on top of the column of seawater, making it stable. Mixing of the water can only occur through diffusion processes related to the water temperature profile rather than its salinity.

5.4 Icebergs

Icebergs are large masses of ice that calve from floating ice shelves or glaciers. Their mass does not contribute to seawater level rise since they were already floating in water. Most of their mass is underwater as their density is lower than that of seawater, which can be shown with the following isostatic equation:

$$\rho_i h = \rho_w h_s, \tag{5.7}$$

where ρ_w is the density of seawater, ρ_i is the density of ice, h is the total iceberg height and h_s is the submerged portion of the iceberg. Substituting the appropriate density indicates that around 90% of an iceberg is submerged.

Antarctic icebergs are given names depending on their formation area following the convention in table 5.1. For example, B15, the largest on record, broke off the Ross Sea Ice Shelf in 2000 and measured over 300 km wide, with a surface area of more than 10 000 km^2. One of the last images available of the B15 iceberg is a daunting, coffin-shape image taken in 2018 (figure 5.14d).

Table 5.1. Naming convention for Antarctic icebergs.

Letter	Longitude	Area
A	0° to 90°W	Bellingshausen Sea, Weddell Sea
B	90°W to 180°	Amundsen Sea, Eastern Ross Sea
C	180° to 90°E	Western Ross Sea, Wilkes Land
D	90°E to 0°	Amery Ice Shelf, Eastern Weddell Sea

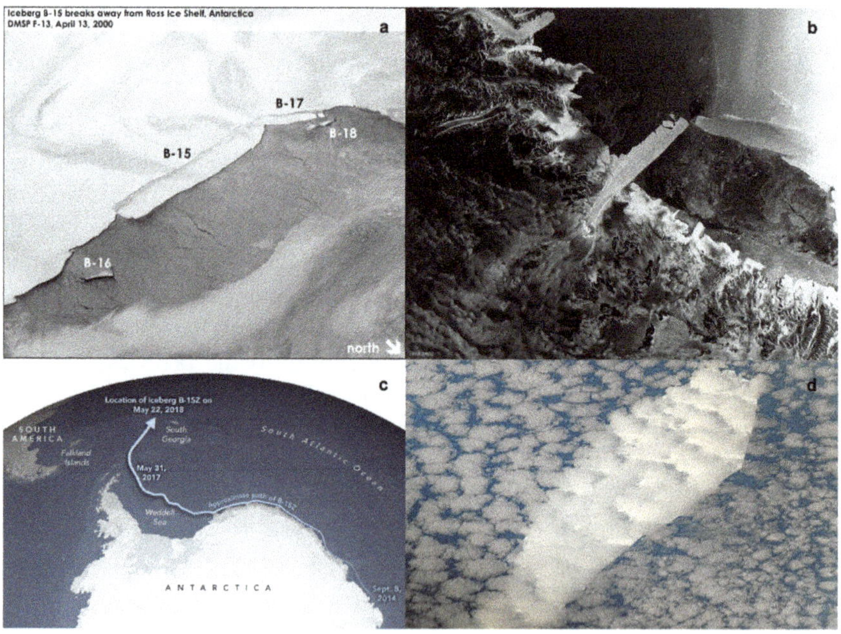

Figure 5.14. The fate of iceberg B15. (a) Satellite image from 2000; (b) B15 touching the Drigalski Ice Tongue; (c) position of B15 in 2018; and (d) aerial image of B15 in 2018. Panel (a) Copyright © University of Wisconsin-Madison, Space Science and Engineering Center, Antarctic Meteorological Research Center; panel (b) courtesy of ESA Earth Watching project, 2005; panels (c) and (d) credit NASA Earth Observatory.

Figure 5.15. The iceberg suspected of sinking the RMS Titanic; a smudge of red paint much like the Titanic's red hull stripe was seen near its base at the waterline. Copyright © Photographer Prinz Adalbert (1912), Uploaded by Mgiganteus, Public Domain.

Icebergs do not just float: green marks can often be seen on the visible part. These are signs of algae, from which it can be deduced that the iceberg must have flipped upside down. Collapse of icebergs can be explosive. This gives quite a spectacular view, though not from that close!

Of course, the most notorious iceberg is the one that caused the sinking of the Titanic during its maiden voyage to New York. Not far from Newfoundland, the ship hit an iceberg, whose sharp submerged edges created a gash along the side of the ship. A large iceberg with traces of red paint was photographed later and the red marks led people to suspect that this was the iceberg responsible for the sinking of the Titanic. It was around 100 m long and 30 m high at the surface (figure 5.15). The sinking of the Titanic led to the establishment of the International Ice Patrol, which aims to monitor and signal the presence of icebergs and collect important meteorological and oceanographic data.

5.5 An overview of our ice sheets: Antarctica and Greenland

5.5.1 Antarctica

Antarctica is the largest ice-covered landmass on Earth. In summer, its size is approximately 13 million square kilometres. Some of the lowest temperatures on Earth were recorded in Antarctica (−89.2 °C at the Soviet Vostok Station on 21 July 1983), although colder temperatures have been measured by satellite. This remote continent has plenty to offer scientists because of its wildness; in fact, many areas of the Antarctic are untouched by civilisation, making it possible to study, for example, our paleoclimate using ice cores. The air bubbles trapped within the ice retain information on the amount of CO_2 present in the atmosphere, giving evidence of global changes in atmospheric temperature. The deepest ice core was drilled not far from Vostok during the EPICA project (Augustin *et al* 2004), measuring over

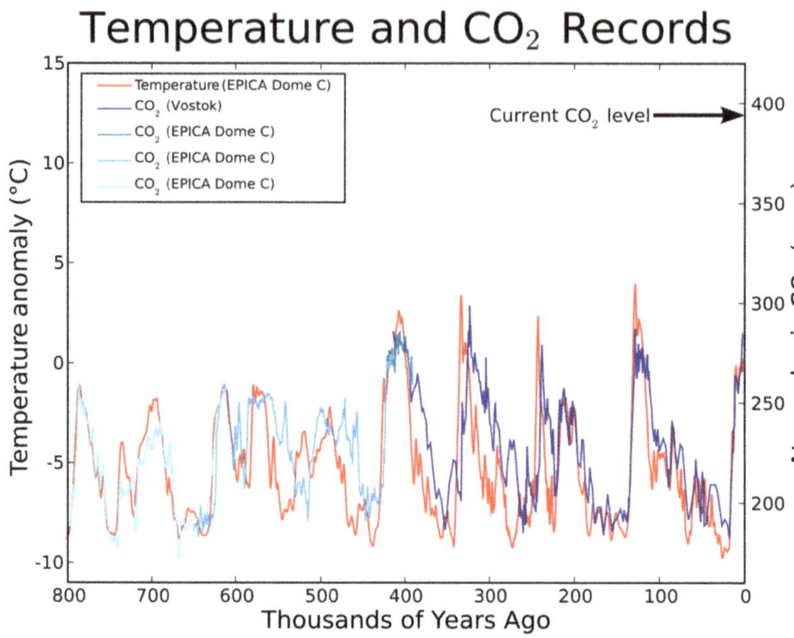

Figure 5.16. The Antarctic temperature changes during the last several glacial/interglacial cycles of the current ice age relative to the current CO_2 level. Red: EPICA Dome C temperature data (Jouzel and Masson-Delmotte 2007). Dark blue: Vostok CO2 data (Petit 2005). Steel blue: EPICA Dome C temperature data, 423–391 kyr BP (Siegenthaler *et al* 2005b). Pale blue: EPICA Dome C CO2 data, 650–413 kyr BP (Siegenthaler *et al.* 2005a). Cyan: EPICA Dome C CO2 data, 800–650 kyr BP (Lüthi *et al* 2008). Copyright © Dr Leland McInnes, Attribution-ShareAlike 3.0 Unported (CC BY-SA 3.0).

3000 m. From this it is possible to observe the past eight glaciations dating back 740 000 years, as shown in figure 5.16.

West Antarctica is partially under sea level because of the weight of the thick ice sheet that covers it, and it is in fact defined as a marine ice sheet. The main reason why ice sheet cover is maintained in this region is the presence of ice bonded to the bedrock, forming buttresses that stop the glaciers' flow. Glacier acceleration is the main cause for Antarctic ice loss (Rignot *et al* 2008). A good example of a fast-receding glacier is Pine Island (figure 5.17), which contributes 25% of the total ice loss from West Antarctica and has now reached a tipping point; it is expected to retreat for tens of kilometres (Favier *et al* 2014). In 2013 an iceberg as large as 700 km^2 detached from the ice shelf, and the one in 5.17 shows the situation in 2020.

5.5.2 Greenland

The Greenland ice sheet is the second largest ice sheet on Earth. It covers the vast majority of Greenland and can be over 3000 m deep in its central part. Its weight pushes down the bedrock so that most of the internal part of the region is below sea level. The coastal glaciers of Greenland have experienced increasing rates of ice loss (figure 5.18). Greenland contributes 0.5 mm yr^{-1} of a total of 3.2 mm yr^{-1} of sea level rise.

Figure 5.17. Landsat image of the Pine Island Glacier ice shelf acquired on 11 February 2020. Image courtesy of NASA. Earth Observatory images by Lauren Dauphin, using Landsat data from the U.S. Geological Survey.

Figure 5.18. Extent of summer melting of the Greenland ice sheet in 1992 and 2002. Image courtesy of NASA, reproduced with kind permission from AMAP, Impacts of a Warming Arctic–Arctic Climate Impact Assessment.

It is widely acknowledged that the Greenland ice sheet has various positive feedback mechanisms related to global warming that lead to increased melting. The temperature has increased by 6°C–8°C. Subsequently, ponding of meltwater on the surface has lowered the albedo and the porosity of the ice sheet has increased, leading to percolation and therefore faster glacier sliding. If Greenland were to melt completely this would cause a sea level rise of about 7 m. In 2012, for the first time, almost the whole of Greenland was subject to surface melting. A 2014 study has also shown an increased melting rate in the northern part of the Greenland ice sheet, which was previously considered to be more stable. This implies that all current forecasts of the disappearance of this ice sheet overestimate the timescale of the process, and it could happen more quickly than expected (Khan *et al* 2014).

References

Augustin L *et al* 2004 Eight glacial cycles from an Antarctic ice core *Nature* **429** 623–8

Bamber J L *et al* 2009 Reassessment of the potential sea-level rise from a collapse of the West Antarctic Ice Sheet *Science* **324** 901–3

Bamber J L, Vaughan D G and Joughin I 2000 Widespread complex flow in the interior of the Antarctic Ice Sheet *Science* **287** 1248–50

Banwell A F, MacAyeal D R and Sergienko O V 2013 Breakup of the Larsen B Ice Shelf triggered by chain reaction drainage of supraglacial lakes *Geophys. Res. Lett.* **40** 2013GL057694

Barnola J-M *et al* 1991 CO_2-climate relationship as deduced from the Vostok ice core: a re-examination based on new measurements and on a re-evaluation of the air dating *Tellus* B **43** 83–90

Favier L *et al* 2014 Retreat of Pine Island Glacier controlled by marine ice-sheet instability *Nature Clim. Change* **4** 117–21

Glasser N F *et al* 2011 Global sea-level contribution from the Patagonian Icefields since the Little Ice Age maximum *Nat. Geosci.* **4** 303–7

Jouzel J and Masson-Delmotte V 2007 EPICA Dome C Ice Core 800 KYr deuterium data and temperature estimates *PANGAEA* (https://doi.org/10.1594/PANGAEA.683655)

Khan S A *et al* 2014 Sustained mass loss of the northeast Greenland ice sheet triggered by regional warming *Nature Clim. Change* **4** 292–9

Ligtenberg S R M, Helsen M M and Van den Broeke M R 2011 An improved semi-empirical model for the densification of Antarctic firn *The Cryosphere Discussions* **5** 1921–48

Lubliner J 2008 *Plasticity Theory* (Mineola, NY: Courier Dover Publications)

Lüthi D *et al* 2008 CO_2 record from the EPICA Dome C 1999 (EDC99) ice core (Antarctica) covering 650 to 800 kyr BP measured at the University of Bern, Switzerland *PANGAEA* (https://doi.org/10.1594/PANGAEA.710901)

Morris E M and Wingham D J 2014 Densification of polar snow: measurements, modeling, and implications for altimetry *J. Geophys. Res. Earth Surf.* **119** 349–65

Petit J-R 2005 Methane and carbon dioxide measured on the Vostok ice core *PANGAEA* (https://doi.org/10.1594/PANGAEA.55501)

Rignot E *et al* 2008 Recent Antarctic ice mass loss from radar interferometry and regional climate modelling *Nat. Geosci.* **1** 106–10

Sandells M and Flocco D 2014 *Introduction to the Physics of the Cryosphere* (San Rafael, CA: Morgan & Claypool)

Siegenthaler U *et al* 2005a EPICA Dome C carbon dioxide concentrations from 423 to 391 kyr BP *PANGAEA* (https://doi.org/10.1594/PANGAEA.472482)
Siegenthaler U *et al* 2005b EPICA Dome C carbon dioxide concentrations from 650 to 413 kyr BP *PANGAEA* (https://doi.org/10.1594/PANGAEA.472481)
Somigliana C 1921 Sulla profondita dei ghiacciai *Rend. R. Acad., Dei Lincei, Cl. d. Sci. Fis.-Mat. Natur* **30** 2
van der Veen C J *et al* 2007 Subglacial topography and geothermal heat flux: potential interactions with drainage of the Greenland ice sheet *Geophys. Res. Lett.* **34** L12501

Melody Sandells and Daniela Flocco

Chapter 6

Sea ice

Sea ice covers the polar oceans for most of the year. Its extent varies noticeably between seasons: its size varies between 6 and 14 million square kilometres in the Arctic and 3 and 19 million square kilometres in the Antarctic. The two poles are very different: the Arctic Ocean is surrounded by land, therefore the maximum size that sea ice can grow is limited and constricted by that land. In contrast, the Antarctic is a continent covered by an ice sheet several kilometres thick and surrounded by oceans, so sea ice growth is not physically constrained.

6.1 Sea ice properties

Sea ice forms from freezing seawater and differs from snow, glaciers and ice sheets because of the presence of salt at its interior. Seawater freezes at a lower temperature than pure water due to its saline content. The decrease in the freezing point of a pure liquid when another substance is dissolved in it is a colligative property of every solution: the lowering of the freezing point is proportional to the number of dissolved particles (molecules or ions), and does not depend on their nature. It is given by

$$\Delta T = K_f C_m, \tag{6.1}$$

where C_m is the molar concentration of dissolved solute and K_f is the cryoscopic constant for the solvent used. Seawater with a salinity of 35 psu[1] freezes at about −1.9 °C. In nature, sea ice is constituted by two parts: salt-free ice and liquid brine, the latter of which is characterised by a very high salt concentration (Weeks and Ackley 1986). Liquid brine pockets are found within the almost pure ice matrix. Given an ion concentration X, the partition constant is defined as $k = \frac{X_{\text{ice}}}{X_{\text{water}}}$.

[1] Practical Salinity Units (psu): corresponds to the ratio between the conductivity of a water sample and the one of a standard solution of potassium chloride with 32.4356 grams of salt in 1 kg of solution at 15 °C. This is a nondimensional measure and 35 psu corresponds to 35 grams of salt in 1 litre of solution.

doi:10.1088/978-0-7503-3647-5ch6

For many solutes in sea ice, k is of the order of 10^{-4}, and therefore sea ice can be considered as a pure phase. From this point of view there are not many differences between freshwater ice and sea ice. The reason for brine pocket formation is that the sodium ions are not as mobile as the hydrogen atoms and the chloride ions do not form hydrogen bonds. Therefore, the liquid brine, which remains within the sea ice, contains all the molecules that are water soluble but insoluble in ice. Furthermore, sea ice only encloses gases at very high pressures. So rather than being solid throughout, sea ice contains liquid brine inclusions, solid salts, microalgae, trace elements, gases and other impurities, which all exist in the interstices of a porous, solid ice matrix (Hunke *et al* 2011).

6.2 Ice formation

From thermodynamics, the ice production P (in units of metres per second) is dependent on the heat transferred from a water body:

$$P = -\frac{Q_{\mathrm{T}}}{\rho_{\mathrm{ice}} L_{\mathrm{H}}}, \tag{6.2}$$

where Q_{T} is the net heat flux, L_{H} is the latent heat of melting of the ice (3.34×10^5 J kg^{-1}) and ρ_{ice} is the density of sea ice (0.95×10^3 kg m^{-3}). Note that sea ice is more dense than pure ice because of its brine content. Ice formation can follow two different processes: a stable (planar) growth, typical of low turbulence areas (e.g. in lakes), or unstable (dendritic[2]) growth, which occurs in the open ocean. A schematic of dendritic growth is represented in figure 6.1 (Gow and Walter 1991).

In the case of planar growth, when air temperature falls below 0 °C, the water on the surface cools down, sinks and activates a convective cell, in which warmer, deeper water rises. By this mechanism the whole water body reaches the temperature of maximum density of water (4 °C). If low air temperatures are maintained a thin layer of ice forms on the surface of the water and a thicker layer slowly builds up.

In the ocean, ice often forms in unstable conditions. Wind and waves act to cool the water, but ice formation requires the presence of nucleation points in the water (e.g. snow flakes, salt crystals). Snow ice forms when a snow flake is enveloped in the water and acts as a nucleating agent. In the absence of nucleation points, seawater can cool to its freezing point (−1.89 °C for a salinity of 34 psu) or below with no ice formation (this is termed *supercooling*). Turbulence may induce a quicker nucleation rate by introducing snow and/or bubbles into the water. Once nuclei are present, ice forms very quickly. Once ice has begun to form then the latent heat produced reverses any supercooling and raises the water temperature until it reaches its freezing point. From this point on, the water absorbs all energy released from ice production.

[2] A crystal that has branched in growth into two parts. Crystals that grow in this way (dendritic growth) have a branching treelike appearance. A Dictionary of Science, Oxford University Press, © Market House Books Ltd 1999.

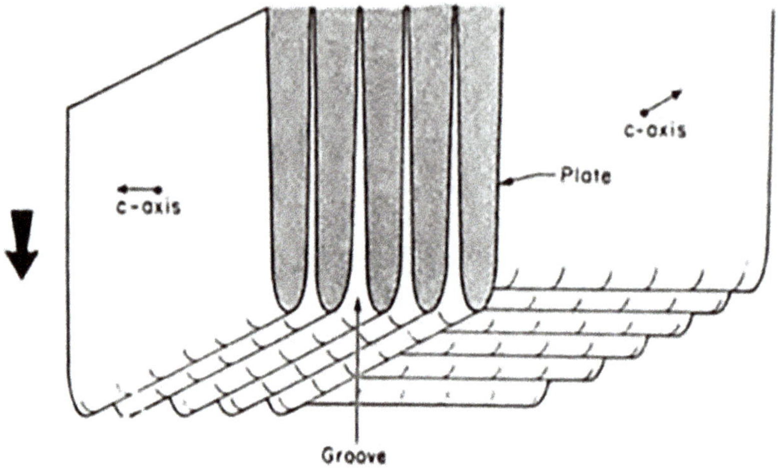

Figure 6.1. Dendritic growth. Image courtesy of Gow and Tucker (1991).

Figure 6.2. Frazil ice on seawater. Copyright © Dany Dumont.

The first ice that forms under turbulent conditions is called frazil ice, described as 'fine spicules or plates of ice in suspension in water' (Martin and Kauffman 1981). Its name comes from the French *fraisil*, meaning cinder (figure 6.2).

If the degree of supercooling is greater than 0.2 °C the frazil crystal grows as a disc without any preferential direction until it reaches a diameter of about 3 mm, following the Bravais law[3]. At this point, the edges of the frazil discs become unstable and are no longer smooth. For even greater supercooling levels the ice

[3] The Bravais rule states that the density of these lattice points, or spacing between lattice planes, is proportional to the relative importance of crystal forms. It follows that the faces most likely to be found on a crystal are those parallel to lattice planes of greatest point density (*A Dictionary of Earth Sciences* © Oxford University Press).

crystals show a preferential growth direction and the growth becomes dendritic for supercooling levels greater than −1 °C. In a turbulent environment such as the ocean, a secondary nucleation process occurs: the first ice crystals behave as new nucleation agents inducing the number of ice crystals to quickly increase.

During dendritic growth, brine can be entrapped in pockets if the ice formation is very fast. The brine drains through gaps between ice stalactites: vertical cylinders of ice almost parallel to each other. It can be observed that the upper layer of a brine pocket is characterised by a lower temperature and higher salinity than the rest of the brine pocket. Brine diffuses towards the bottom of the ice stalactite, following the negative concentration gradient, and it is replaced by lower-salinity water. The overall decrease of salinity in the ice is described as follows:

$$S = S_0 \left(\frac{T_{i_0}}{T_i} \right)^{\frac{\Delta\rho}{1-\Delta\rho}}, \tag{6.3}$$

where S_0 is the initial salinity and $\Delta\rho$ (~0.1) is the density difference between water when the ice temperature changes from T_i to T_{i_0} (Weeks and Ackley 1986). Knowledge of the bulk salinity of the ice gives an indication of the age of the ice: old ice has a lower bulk salinity than young ice. Note that this is not because older ice has had more time to desalinate but rather because a larger fraction of the ice grew at a lower rate in the older ice, which reduces bulk salinity (Petrich *et al* 2011).

Calculations of sea ice salinity can also be made from the open water salinity using the following formula derived by an empirical approach: Martin and Kauffman (1981) show in a laboratory experiment that in high-turbulence areas the ratio of ice to open water salinity is almost constant:

$$\frac{S_\mathrm{i}}{S_\mathrm{w}} = 0.31. \tag{6.4}$$

This approach is valid for frazil ice and grease ice production (see section 6.3), but may be too simplistic for other forms of ice. Initially, sea ice grows preferentially by area rather than thickness. When the sea ice coalesces and forms platelets, they attain a typical pancake shape characterised by rough raised edges due to the friction between ice platelets. The diameter of the pancakes varies from a few centimetres to over 2 metres. The pancakes grow closer to each other, forming one big expanse of ice that is the base for ice-pack formation and thickening of the sea ice (figure 6.3).

6.3 Sea ice classification

Sea ice can be classified in terms of its age, its crystallographic characteristics and its location. The term *new ice* indicates recently formed ice. The principal characteristic of new ice is that the ice crystals are not compacted and do not form a lattice. Frazil ice, grease ice and nilas are included in this category, as shown in table 6.1. Frazil ice forms in various ways, but the turbulence produced by waves and wind is the major reason for the ice crystals not to consolidate.

Figure 6.3. Pancake ice in the Drake Passage, Southern Ocean. Photo courtesy of Kenneth D Mankoff. This photo is distributed by the EGU under a Creative Commons licence.

Table 6.1. New ice classification. See ASPeCt Glossary for more information.

Ice	Description
Frazil ice	Fine spicules or plates of ice suspended in water. Frazil crystals are usually suspended in the top few centimetres of the ocean's surface and give the water an oily appearance.
Grease ice	A later stage of freezing than frazil ice when the crystals have coagulated to form a 'soupy' layer on the surface.
Nilas	A thin elastic crust of ice easily bending on waves and swell. It has a matt surface and is up to 10 cm in thickness.

One way in which frazil ice can form is by adiabatic supercooling: if a body of water rises to the surface from a significant depth, the pressure diminishes, its freezing point rises and thus ice forms. This behaviour is described by the aforementioned Clausius–Clapeyron equation (see also equation (2.3)):

$$\frac{dP}{dT} = \frac{L}{T\Delta v}, \tag{6.5}$$

where P is the pressure, T the temperature, L the latent heat of fusion for water and Δv the specific volume change during the phase change. A phase diagram of water with respect to temperature and pressure is shown in chapter 2, figure 2.3.

Once pancakes coalesce, a layer of new ice covers the water surface leading to congelation growth: this is typical of ice growing from existing sea ice and characterised by a columnar crystal texture. As previously mentioned, sea ice can be classified according to its location and its interactions with the surrounding environment. The area of the polar sea ice cover which remains ice covered in winter but is ice-free during the summer is defined as the seasonal sea ice zone. Sea ice that interacts with the open ocean and is affected by the influence of waves is defined as the marginal ice zone and must also be regarded as part of the seasonal sea ice zone (Wadhams 1986). In terms of its persistence, sea ice can be distinguished between first year ice, which forms every winter from freezing seawater, and multiyear ice, which is ice surviving at least a polar summer.

A systematic way to describe ice observation, as used in the field, is to fill ice egg charts: figure 6.4 gives an example and explanation of the various symbols.

The 'egg code' is divided into four sections:

1. Total concentration (C_t): the ice coverage of an area determined by its concentration and expressed in tenths.
2. Partial concentration (C_a, C_b, C_c): the break-down of the total ice coverage expressed in tenths and graded by thickness (thickest-2a, medium-2b, thinnest-2c.) These grades directly relate to the type of ice described in the egg chart section 3.
3. Stage of development (S_a, S_b, S_c): the type of ice in each of the grades, in sections 2a, 2b and 2c above, determined by its age (old ice or young ice) and expressed as a number.
4. Floe size (F_a, F_b, F_c): the form of the ice determined by its floe size (any relatively flat piece of ice 20 m or more across) for each section and expressed as a code number.

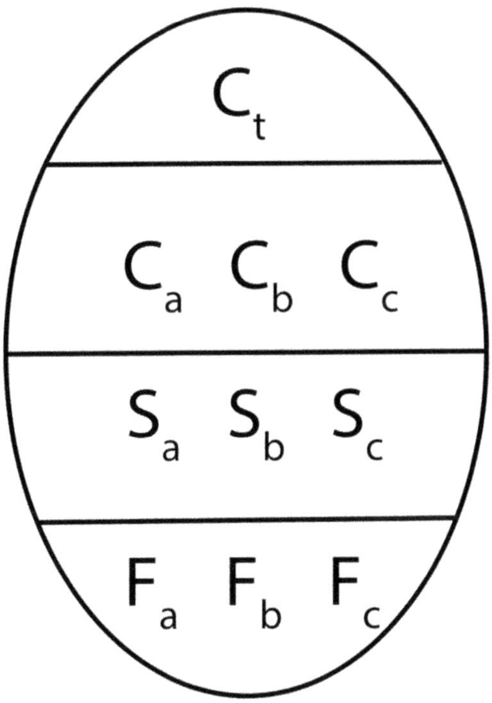

Figure 6.4. Egg code.

6.4 Sea ice modelling

More observations of sea ice are needed in order to improve the efficacy of model predictions, but expeditions to the poles are expensive and fieldwork can prove to be hazardous. For many months per year the polar night makes it more difficult both to reach these remote areas and to carry out field experiments. Satellite observations have greatly improved the amount of data available over large areas and at a high temporal frequency, although every satellite measurement technique needs to be validated before the data are released for use. Available observations are used to improve the physics of sea ice models and to test and validate their accuracy. Models can then be used to study the interactions between the polar areas and changes in the climate.

The behaviour of sea ice is modelled using equations that govern three different aspects of the sea ice physics:

- Momentum balance: the ice reactions to wind, currents and internal sea ice interactions.
- Energy balance: concerning the thermodynamic aspect of sea ice behaviour, describes the heat exchanges between the ice, the atmosphere and the ocean, thus determining the sea ice growth and melt.
- Mass balance: exchanges between phases, flow and precipitation all contribute to the amount of sea ice present.

The momentum balance is described by the following equation:

$$m\frac{D\mathbf{u}}{Dt} = \tau_a + \tau_w + \nabla \cdot \sigma - mf_c\mathbf{k} \times \mathbf{v} - mg\nabla \cdot H, \tag{6.6}$$

where m is the ice mass and $\mathbf{u}$ is the ice drift. The dominant forces are air drag (τ_a), ocean drag (τ_w) and the ice–ice force ($\nabla \cdot \sigma$), with σ representing the stress caused by rupture, frictional sliding, pressure ridging and collisions. The internal ice stress has direct consequences on the ice drift, as the response of ice to external forces such as wind and currents will be impacted by its strength. The other two terms represent the Coriolis force due to planetary rotation and the gravity force from sea surface tilt ($\nabla \cdot H$).

The energy balance of sea ice at the sea ice–atmosphere interface is described by the following equation:

$$F_{LW} - \varepsilon\sigma T^4 = F_{SW}(1 - \alpha)(I - I_0) - F_{sens} - F_{lat} = k_{eff}\frac{\partial T}{\partial z}, \tag{6.7}$$

where F_{LW} and F_{SW} are the downwards fluxes of longwave and shortwave radiation, $\varepsilon = 0.99$ is the thermal emissivity of sea ice, $\sigma = 5.67 \times 10^{-8}$ J (K^4 m^2 s)$^{-1}$ is the Stefan–Boltzmann constant, α is the albedo, I_0 is the fraction of solar radiation penetrating the surface layer (considered to be of negligible thickness), and F_{sens} and F_{lat} are the sensible and latent-heat fluxes (Maykut and Untersteiner 1971, Ebert and Curry 1993, Taylor and Feltham 2004, Bailey *et al* 2010).

For conservation of mass, the sea ice thickness is represented by a statistical distribution, $g(\mathbf{x}, h, t)$, such that

$$\int_0^{\infty} \frac{\partial g}{\partial t} = 1. \tag{6.8}$$

The evolution of $g(\mathbf{x}, h, t)$ is given by

$$\frac{\partial g}{\partial t} = -\nabla \cdot (gu) - \frac{\partial}{\partial h}(fg) + \psi, \tag{6.9}$$

where u is the horizontal velocity, f is the rate of thermodynamic ice growth and ψ is a ridging redistribution function.

One of the main aspects of sea ice modelling is the parameterisation of the redistribution function ψ that governs the statistical behaviour of sea ice. Any redistribution function has to satisfy a conservation rule. The ice–ice force, given by the rheology (the study of the internal forces of sea ice) of sea ice, is poorly understood; therefore this aspect is still the subject of research investigations.

The ice growth is described by the Stefan approximation that links the ice growth to the temperature difference between the ice surface and the ice bottom:

$$\frac{\rho_{\text{ocean}}}{\rho_{\text{ice}}} L\phi \frac{\partial h_{\text{ocean}}}{\partial t} = k_{\text{eff}} \frac{\partial T}{\partial z} + F_{\text{ocean}}, \tag{6.10}$$

where ρ_{ocean} and ρ_{ice} are, respectively, the densities of the seawater and the sea ice, ϕ is the solid fraction of sea ice at the ice–ocean interface and F_{ocean} is the heat flux at the sea ice bottom (Bitz and Lipscomb 1999, Feltham *et al* 2006, Bailey *et al* 2010).

Inside, the sea ice can be considered a porous medium characterised by a rigid matrix of almost pure ice with interstices filled by liquid brine. The evolution temperature in the ice interior is described by the mushy layer equation, which can be written in the following terms (Feltham *et al* 2006):

$$c_{\text{eff}_{\text{ice}}} \frac{\partial T}{\partial t} = \frac{\partial}{\partial z}\left(k_{\text{eff}_{\text{ice}}} \frac{\partial T}{\partial z}\right) + A_r, \tag{6.11}$$

with the phase change terms held implicitly within the effective heat capacity and conductivity, and A_r being the absorption of solar radiation that penetrates through the upper sea ice surface. Feltham *et al* (2006) demonstrate the equivalence of the mushy layer equation with the diffusion equation (Maykut and Untersteiner 1971), but accounting for the phase change by calculating an effective specific capacity and conductivity of sea ice c_{eff} and k_{eff}.

6.5 The importance of albedo: melt ponds, leads and polynyas

The polar ice caps on sea are not uninterrupted expanses of ice. Indeed, they exhibit many discontinuities such as *leads* and *polynyas*. This affects the albedo of the sea ice surface. The albedo is the characteristic of a body to reflect light: as described in chapter 2, it is defined as the ratio between the light absorbed by a surface and the total light that reaches the surface.

A dark surface absorbs more energy than a lighter one and therefore warms up more quickly. With the disappearance of sea ice and snow two effects have been observed in the high latitudes: the tree line in the Northern Hemisphere has moved to lower latitude, leaving darker surfaces uncovered; and with the sea ice disappearance over the Arctic, more ocean surface is exposed to the Sun. Over the Arctic this process is of particular importance because warming of the surface directly enhances the melting rate of sea ice, while the warming up of the ocean leads to further melting: this process is known as the albedo feedback mechanism.

6.5.1 Polynyas

Polynyas are areas that remain ice-free during the polar winter (figure 6.5). There are two different processes that can lead to the formation of a polynya. In the first case, an upwelling of warm water prevents the surface from freezing. Occuring far from

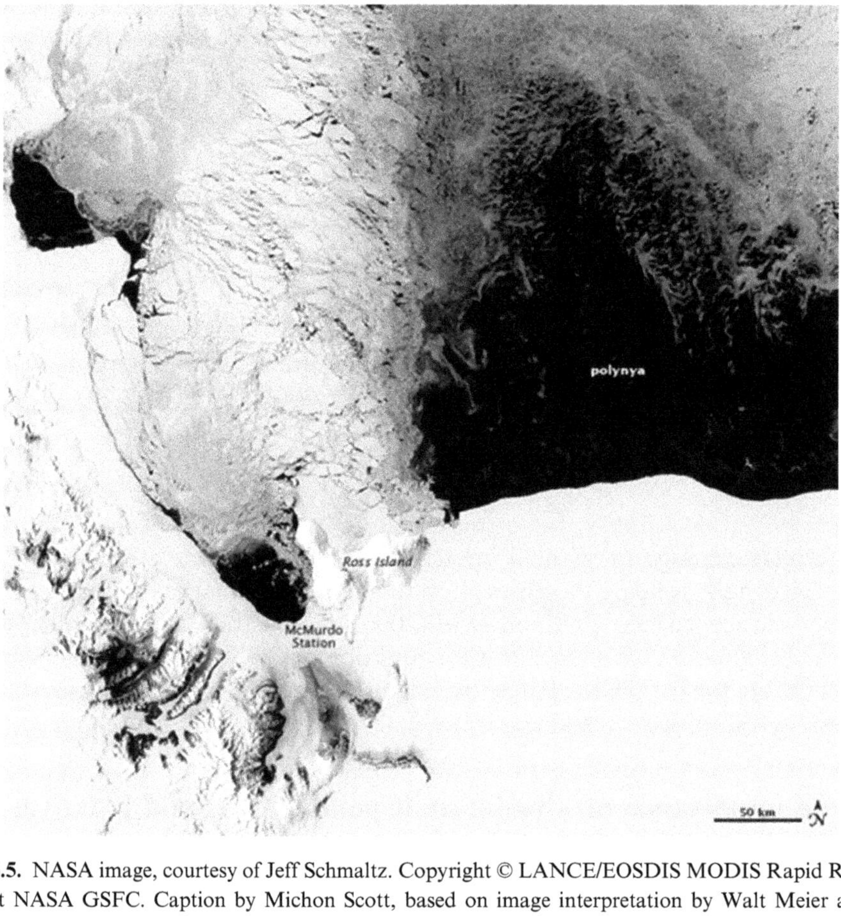

Figure 6.5. NASA image, courtesy of Jeff Schmaltz. Copyright © LANCE/EOSDIS MODIS Rapid Response Team at NASA GSFC. Caption by Michon Scott, based on image interpretation by Walt Meier and Ted Scambos, National Snow and Ice Data Center.

the coast, these are also called *open ocean* polynyas. Another common process in the Antarctic region is the quick removal of ice by strong katabatic winds blowing from the continent, and so formation close to the coast. In this case we observe the formation of a *latent-heat* polynya, or *coastal* polynya. The two processes described are not mutually exclusive, so that every polynya can present a mixed character.

Latent-heat polynyas play a fundamental role in dense water production. Frazil ice is continuously formed and removed by the wind thereby leaving the area ice-free. The large difference in temperature between the cold air and warm water leads to a strong heat loss from the sea surface, which freezes almost immediately, activating as a positive feedback (Fusco *et al* 2002). The ice production rate increases heavily, leading to a large release of brine in the column of water underneath, strongly modifying its structure. The densest waters of the Southern Ocean are formed in polynya areas. In particular, all these processes are typical of an area where a broad continental margin is present. The brine formed in the ice formation process has the time to reside locally and contribute to the salinification of the basin.

Estimation of a polynya's surface area allows the evaluation of the amount of salt, and hence dense water, produced in these areas, which is needed in order to describe the dynamics of the column of water. Polynyas are fairly consistent features: their formation is very often due to recurrent meteorological patterns. Forecasting of their opening and maintenance is therefore possible.

The polynya dimension is defined as the distance between the zone where frazil ice forms and the one where it collects against the offshore pack ice (Pease 1987). The frazil ice moves more quickly than the pack ice even though they are subject to the same wind speed. Frazil ice is therefore compacted against the pack ice downwind (figure 6.6). From a dynamic point of view, a polynya remains open, reaching its equilibrium if the volume of ice produced in the polynya is equal to or is exceeded by the volume transported from the polynya elsewhere. In the Pease flux model the ice

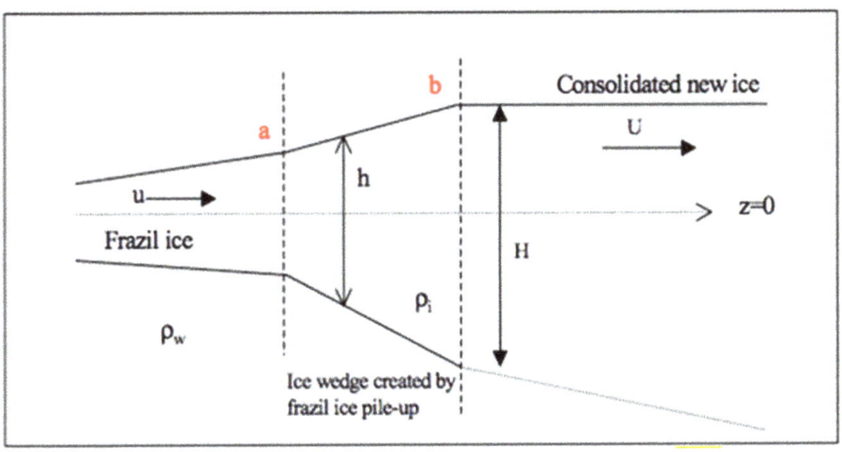

Figure 6.6. Schematic of the frazil ice wedge formed at the polynya edge. Image courtesy of Biggs *et al* (2000), Copyright, with permission from Cambridge University Press.

rheology is not taken into account and the speed at which the ice collects is set as infinite: once it forms, ice is immediately pushed towards the edge of the polynya.

Further studies have implemented the Pease model and developed a two-dimensional flux model (figure 6.7) that better accounts for the conservation of frazil ice mass during the frazil accumulation process (Biggs and Willmott 2004, Biggs *et al* 2000).

Ice can be removed by wind, currents or by their combined action, and it is observed to concentrate in long strikes parallel to the wind direction due to an underlying Langmuir circulation. Langmuir cells form when a differential wind speed along a coast induces the formation of small wind eddies in the downwind direction. The vorticity of the rings provokes the formation of convective cells with alternate lines of convergence and divergence, characterised by downwelling and upwelling processes, respectively (figure 6.8). Ice collects in the downwelling streaks and it is brought elsewhere in a direction parallel to the wind, as shown in figure 6.8.

Two main modelling approaches are used to describe polynya dynamics: flux models and dynamic-thermodynamic models. Essentially they differ in the way in which they parameterise the ice behaviour. In flux models, a very important but poorly understood process is the frazil ice-collection depth (*H*) at the downwind edge of the polynya (Winsor and Björk 2000). In their laboratory study, Martin and Kauffman (1981) define the line where grease ice collects and starts behaving as a solid body as the 'dead line'. It is difficult to extend this definition to a real environment: the ice-collection depth depends on many variables such as wind speed, air temperature and fetch. A laboratory experiment carried out by the Scott

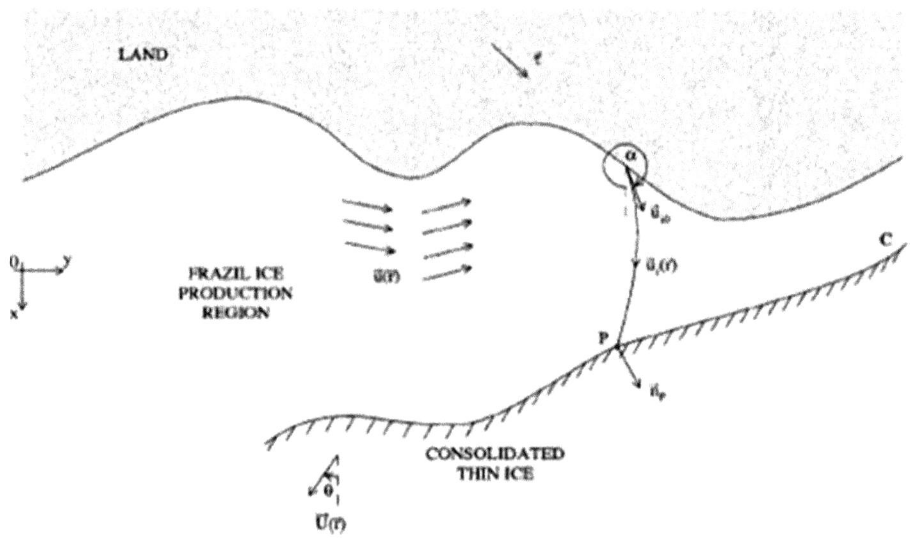

Figure 6.7. Schematic of the two-dimensional polynya modelling approach: any point *P* on the polynya ice edge *C* is determined so that the fluxes of frazil ice and consolidated ice balance across *C*. *t*, *u*, *ui* and *U* denote the wind stress, the surface velocity of the ocean, the frazil ice velocity and the consolidated thin ice velocity, respectively. In general, all these fields depend on position. Also shown is the angle *a* between the initial point of a frazil ice trajectory and the *x*-axis and the angle *u* between the consolidated thin ice flux and the *x*-axis. Image courtesy of Willmott *et al* (2007).

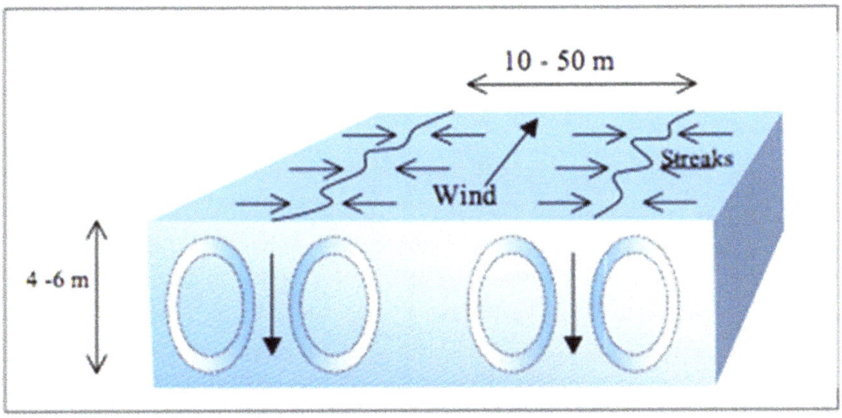

Figure 6.8. Langmuir circulation schematics.

Polar Research Institute in the framework of the European project INTERICE-3, in Hamburg, confirmed the existence of a wedge where pancakes pile up when the kinetic energy of wind and waves, used as energy input, drops down (P Wadhams, private communication, 2001).

In a dynamic-thermodynamic approach the processes regarding the internal ice stresses are taken into account. No difference is made between different types of ice to define the polynya's edges. From this point of view the polynya's edge cannot be clearly indicated. Furthermore, because the dynamic-thermodynamic models are fully computational, the ice drift along the polynya can more readily be assumed variable (Bjornsson *et al* 2001).

6.5.2 Leads

Leads are long cracks in the sea ice pack that form from divergences in pack ice due to the action of currents and wind. Their width can vary between tens of metres to kilometres (figure 6.9). Unlike polynyas, they do not form in recurring locations but behave in a similar way regarding accelerated sea ice formation when water is exposed to cold air.

Leads are important for sea ice dynamics because once they start refreezing they are covered by ice that is thinner and therefore weaker than the surrounding pack ice.

6.5.3 Melt ponds

Melt ponds develop over Arctic sea ice during the melting season from the accumulation of meltwater from ice and snow. These have become increasingly important over the last few decades because they have been more prevalent and absorb much more solar energy due to their dark colour compared to the highly reflective white sea ice. Where ponds form, the ice beneath becomes thinner due to increased melting. Towards the end of the summer, the air temperature lowers and a thin layer of ice forms over melt ponds. Pond meltwater trapped in the ice acts as a storage of heat and does not allow the underlying ice to start thickening until all the

Figure 6.9. Ice regrows in a lead between two sea ice floes in the Beaufort Sea, Arctic Ocean, in March 2013. Image courtesy of S L Farrell.

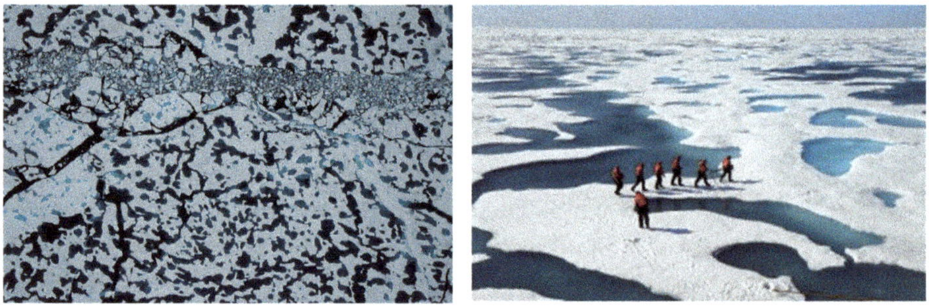

Figure 6.10. Melt ponds litter the surface of Arctic sea ice. Left image credit: Don Perovich, Sea ice researchers, surrounded by Arctic melt ponds, set out to check the thickness of an ice floe. Right image credit: Don Perovich.

pond's water is frozen. Such ponds are up to 1.5 m deep and it can take over 2 months to freeze their volume of water. Considering that ponds cover up to 50% of the sea ice extent, their impact cannot be neglected (figure 6.10).

A strong negative correlation exists between the change in successive mean winter ice thickness and the length of the intervening melt season, suggesting that summer melt processes play a dominant role in determining mean Arctic sea ice thickness for the following winter. Another hint for the importance of melt ponds towards explaining thinning of sea ice is that melt ponds are present in the Arctic more than in the Antarctic, where the sea ice thinning is less striking.

Ponds are rather irregular in shape but occur in larger percentage over thin, young ice: since the area of young ice is increasing (relative to the total amount of ice, which is instead decreasing), the melt ponds' impact will become increasingly important too. This will lead to a positive feedback effect in which thin ice will start thickening later in winter and will possibly be a preferential area for melt ponds' formation in the following spring. Furthermore, in correspondence to where melt ponds form, specular lenses of freshwater form under the sea ice cover, impacting the freezing point of water at the ice–ocean interface. At the beginning of the season sea ice is impermeable, so once ponds form they can be above sea level. When the ponds start melting the ice, it becomes more permeable. When the ponds are fully developed they are in hydrostatic balance[4] with the ocean so they drain to sea level.

Schemes handling melt ponds have only recently been included in global circulation models and are rather crude: the meltwater was assumed to be flushed into the ocean without dwelling on the sea ice. Lack of a melt pond parameterisation can give an overestimate of sea ice thickness of up to 40% during summer (Flocco *et al* 2010, Flocco and Schroeder *et al* 2012). Model results have also shown a good ability to forecast the minimum September ice extent, relating it to the melt pond area calculated by the model in May (Schröder *et al* 2014).

Refreezing of melt ponds also has the effect of delaying sea growth at the end of summer (Flocco *et al* 2015), therefore contributing even further to the Arctic sea ice thinning. The relevance of reproducing the physics of melt ponds in general circulation models has been further confirmed by a recent study on the Last Interglacial (LIG). The LIG is a warmer period 130–116 ka before the present and has the potential to reproduce a climatic scenario similar to the one expected in the future, with Arctic land summer temperatures 4 °C–5 °C higher than the preindustrial era (Guarino *et al* 2020). This study also confirmed that sea ice loss has been the driver of the higher temperature, and suggested that melt ponds might have a relevant role in this process. Further studies (Diamond *et al* 2021) have supported this hypothesis, and confirmed that the positive albedo feedback mechanism was strongly enhanced by the presence of melt ponds and contributed to surface albedo decline, leading to ice-free summers.

References

Bailey E, Feltham D L and Sammonds P R 2010 A model for the consolidation of rafted sea ice *J. Geophys. Res. Oceans* **115** C04015

Biggs N R T and Willmott A J 2004 Unsteady polynya flux model solutions incorporating a parameterization for the collection thickness of consolidated new ice *Ocean Model.* **7** 343–61

Biggs N R T, Maqueda M A M and Willmott A J 2000 Polynya flux model solutions incorporating a parameterization for the collection thickness of consolidated new ice *J. Fluid Mech.* **408** 179–204

Bitz C M and Lipscomb W H 1999 An energy-conserving thermodynamic model of sea ice *J. Geophys. Res. Oceans* **104** 15669–77

[4] A fluid is in hydrostatic balance when it is at rest.

Bjornsson H *et al* 2001 Polynyas in a high-resolution dynamic-thermodynamic sea ice model and their parameterization using flux models *Tellus A* **53** 245–65

Diamond R *et al* 2021 The contribution of melt ponds to enhanced Arctic sea-ice melt during the Last Interglacial *Cryosphere* **15** 5099–114

Ebert E E and Curry J A 1993 An intermediate one-dimensional thermodynamic sea ice model for investigating ice-atmosphere interactions *J. Geophys. Res. Oceans* **98** 10085–109

Feltham D L *et al* 2006 Sea ice is a mushy layer *Geophys. Res. Lett.* **33** L14501

Flocco D, Feltham D L and Bailey E *et al* 2015 The refreezing of melt ponds on Arctic sea ice *J. Geophys. Res. Oceans* **120** 647–59

Flocco D, Feltham D F and Turner A K 2010 Incorporation of a physically based melt pond scheme into the sea ice component of a climate model *J. Geophys. Res. Oceans* **115** C08012

Flocco D and Schroeder D *et al* 2012 Impact of melt ponds on Arctic sea ice simulations from 1990 to 2007 *J. Geophys. Res. Oceans* **117** C09032

Fusco G *et al* 2002 Dynamics and variability of Terra Nova Bay polynya *Mar. Ecol.* **23** 201–9

Gow A J and Tucker III W B 1991 Physical and dynamic properties of sea ice in the polar oceans *Technical Report* DTIC

Guarino M-V *et al* 2020 Sea-ice-free Arctic during the Last Interglacial supports fast future loss *Nat. Clim. Chang.* **10** 928–32

Hunke E C *et al* 2011 The multiphase physics of sea ice: a review for model developers *Cryosphere* **5** 989–1009

Martin S and Kauffman P 1981 A field and laboratory study of wave damping by Grease Ice *J. Glaciol.* **27** 283–313

Maykut G A and Untersteiner N 1971 Some results from a time-dependent thermodynamic model of sea ice *J. Geophys. Res. (1896–1977)* **76** 1550–75

Pease C H 1987 The size of wind-driven coastal polynyas *J. Geophys. Res.* **92** 7049–59

Petrich C, Langhorne P and Eicken H 2011 Modelled bulk salinity of growing first-year sea ice and implications for ice properties in spring *Proc. 21st Int. Conf. on Port and Ocean Engineering under Arctic Conditions* pp 10–4

Schröder D *et al* 2014 Arctic sea-ice minimum predicted by spring melt-pond fraction *Nature Clim. Change* **4** 353–7

Taylor P D and Feltham D L 2004 A model of melt pond evolution on sea ice *J. Geophys. Res. Oceans* **109** C12007

Wadhams P 1986 The seasonal ice zone *The Geophysics of Sea IceNATO ASI Series* ed Untersteiner N (New York: Springer) pp 825–991

Weeks W F and Ackley S F 1986 The growth, structure, and properties of sea ice *The Geophysics of Sea IceNATO ASI Series* ed Untersteiner N (New York: Springer) pp 9–164

Willmott A J, Holland D M and Maqueda M A M 2007 Polynya modelling *Elsevier Oceanography Series* vol 74 (Amsterdam: Elsevier) pp 87–125

Winsor P and Björk G 2000 Polynya activity in the Arctic Ocean from 1958 to 1997 *J. Geophys. Res.* **105** 8789–803

IOP Publishing

Introduction to the Physics of the Cryosphere (Second Edition)

Melody Sandells and Daniela Flocco

Chapter 7

High-latitude processes

7.1 Oceans

7.1.1 The importance of high-latitude processes

In this primer, we are concentrating on the cryosphere and the processes that happen at high latitudes, but it is also critically important to understand that the world's oceans are all interconnected and the energy balance of the Earth depends on phenomena that happen far away. In this chapter, we focus on the oceanography at high latitudes and show how processes that occur there influence the mid-latitude climate.

Oceanic water covers about 71% of the Earth's surface and contains up to 97% of the Earth's water. Historically, four oceans have been identified: the Atlantic, Pacific, Indian and Arctic Oceans. However, in 2000, the International Hydrographic Organization recognised a fifth ocean: the Southern Ocean, stretching from the coasts of Antarctica to 60°S.

But why do we focus our attention to the study of the *polar* waters? Here are a few important points worth outlining:

- The masses of water that are formed here trigger the global thermohaline circulation and therefore the redistribution of heat.
- They intervene in and condition the energy balance of the planet.
- The circulation in the North Atlantic is crucial for the climatic balance of the northern sector of the Northern Hemisphere.
- The Southern Ocean, through the Antarctic Circumpolar Current, transports and redistributes the main physical, chemical and biological properties between all ocean basins.
- The poles are the largest reserve of CO_2.

7.1.2 The Atlantic meridional overturning circulation within the global thermohaline circulation or great conveyor belt

The Sun's energy is not uniformly distributed to the Earth's surface due to the Earth's orbit around the Sun and to the inclination of its axis, therefore a heat surplus is

doi:10.1088/978-0-7503-3647-5ch7

observed at the equator and a heat deficit at the poles. This imbalance drives the observed global atmospheric and oceanic circulations. The large-scale ocean circulation, driven by differences in the salinity and temperature (and therefore density) of water, is defined as the thermohaline circulation. Climate scientist Wallace Smith Broecker introduced the concept of the global ocean conveyor belt, defined by the National Oceanic and Atmospheric Administration (NOAA) as a 'constantly moving system of deep-ocean circulation driven by temperature and salinity'.

The ocean currents, the movement of the ocean in the surface layer, are driven mostly by wind, but the activation of deeper current is due to the formation of dense waters that sink in the deep ocean, inducing what is called an overturning circulation. The polar oceans are preferential locations of dense-water formation: these form during winter when the surface of the ocean interacts with the cold polar atmosphere in two ways:

- They cool down by loss of heat and/or evaporation.
- Their salinity increases by evaporation and ice formation (brine rejection).

In these regions, in fact, surface waters are often at their freezing point because they are in contact with the cold air and become denser by cooling. If the air temperature decreases further, sea ice formation and consequent brine rejection takes place. The brine entrainment leads to the formation of salty dense water that sinks to the ocean depths. Sinking of surface water into the ocean forces the deep water to move horizontally until it finds an area on the planet where it can rise back to the surface and close the current loop. Upwelling usually occurs in the equatorial ocean, mostly in the Pacific and Indian oceans.

The Atlantic Ocean is characterised by a large-scale process defined as the Atlantic meridional overturning circulation (AMOC), which will here be explained in a simplified manner. The AMOC shown in figure 7.1 is a system of surface and deep current through which are shown the interactions between cold, dense water forming at the poles and salty, warm water moving from the equator. The Gulf Stream is the most important current of the Atlantic Ocean and plays a fundamental role in the AMOC: it is a warm surface current originating in the Gulf of Mexico and flows along the coasts of Florida, reaching Newfoundland in Canada to then cross the Atlantic. At this point the Gulf Stream splits in two branches: one flows northward and the other southwards, recirculating water again towards the Gulf of Mexico. The northward branch flows to the Arctic Ocean bordering the coasts of Northern Europe, influencing its climate and mitigating against its northern latitude. Once the Atlantic, salty water of the Gulf Stream meets the cold, fresher Arctic water, it sinks below it, filling part of the Arctic basin.

The Arctic is the location where the so-called North Atlantic Deep Water (NADW) forms. The NADW is a rather complex ensemble of dense waters characterised by temperatures ranging between 2 °C and 4 °C and salinities of 34.9–35.0 Practical Salinity Units (psu). A fraction of the NADW has its origins in the Labrador Sea through deep convection and is characterised by low saturation in chlorofluorocarbon (CFC) levels (Smethie *et al* 2000) while the other location of

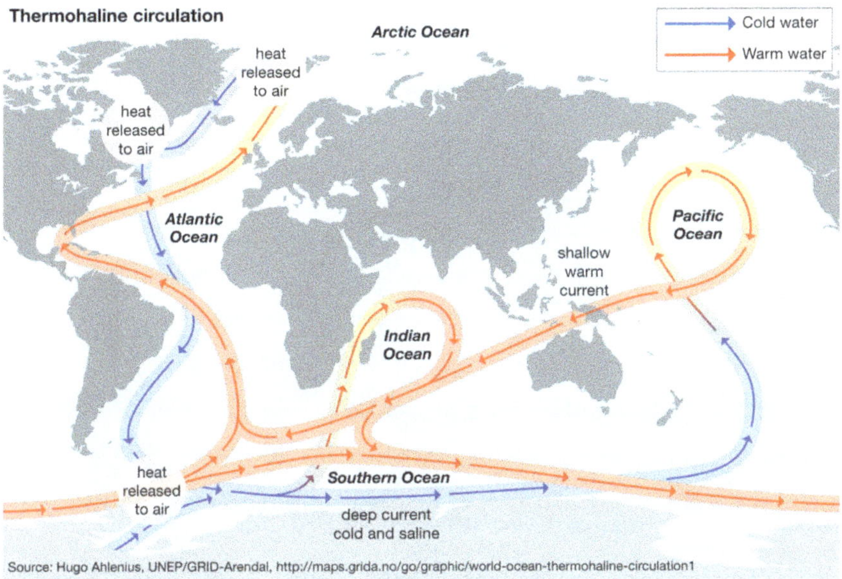

Figure 7.1. Atlantic Meridional Overturning Circulation (AMOC). Image distributed courtesy of Hugo Ahlenius, UNEP/GRID-Arendal.

formation of the NADW is the north of Greenland, where the mechanism of formation of deepwater masses is linked to sea ice processes (Rudels and Quadfasel 1991). In the Southern Hemisphere, the main circulation is driven by the presence of the Benguela current: this is a broad, northward flowing ocean current that forms the eastern portion of the South Atlantic Gyre. The Benguela current plays a key role for the AMOC by transporting heat and salt from the Indian Ocean. Eddies from the warm South Indian Ocean Agulhas current along South Africa's east coast come round the Cape of Good Hope from time to time to join the Benguela current. In the Southern Ocean, along the Antarctic coasts, a dense, cold and salty body of water named the Antarctic Bottom Water (AABW) sinks, filling the depths of the Atlantic ocean. Figure 7.2 shows the relationship between the AMOC and the Great Conveyor Belt.

By looking at a latitudinal section of the Atlantic, as shown in figure 7.3, we can observe the interactions between the bodies of water described and their interactions. This complex system is defined as the aforementioned AMOC. An equivalent overturning circulation is not observed in the Pacific. In fact, deepwater formation may have last occurred in the North Pacific Ocean during the Pliocene, enabling a Pacific meridional overturning circulation that is absent in today's ocean (Thomas *et al* 2021).

7.1.3 Deep convection

Sir Benjamin Thompson, in 1797, described one of the first observations of the NADW at depth at low latitudes: 'It appears to be extremely difficult, if not quite impossible, to account for this degree of cold at the bottom of the sea in the torrid

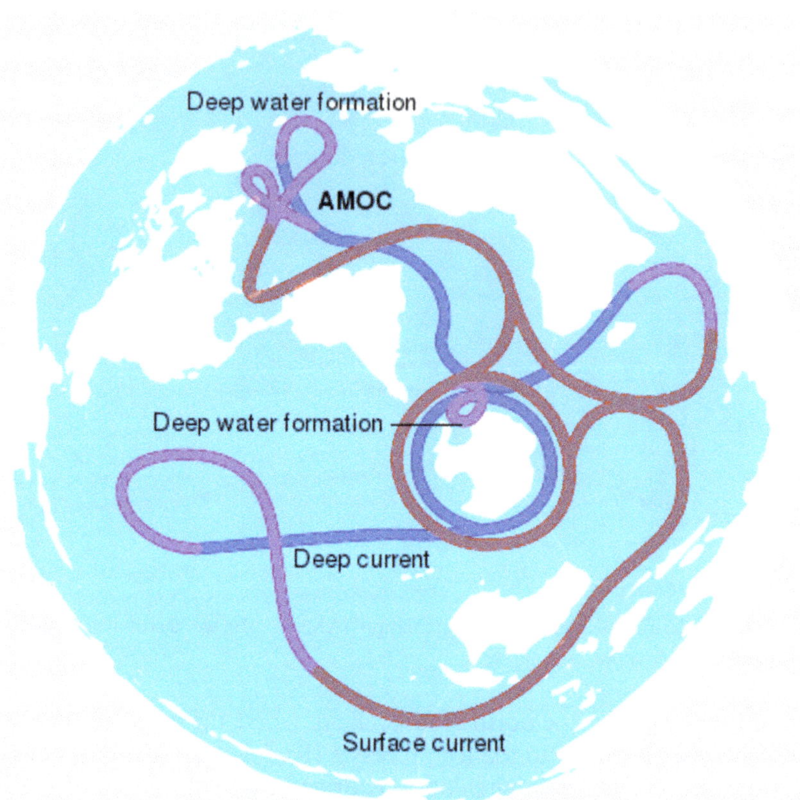

Figure 7.2. Schematics of the Atlantic Meridional Overturning Circulation. Image by CMG Lee, redistributed under the CC BY-SA 4.0 license. Copyright © C A Vreugdenhil and B Gayen, Attribution 4.0 International (CC BY 4.0)

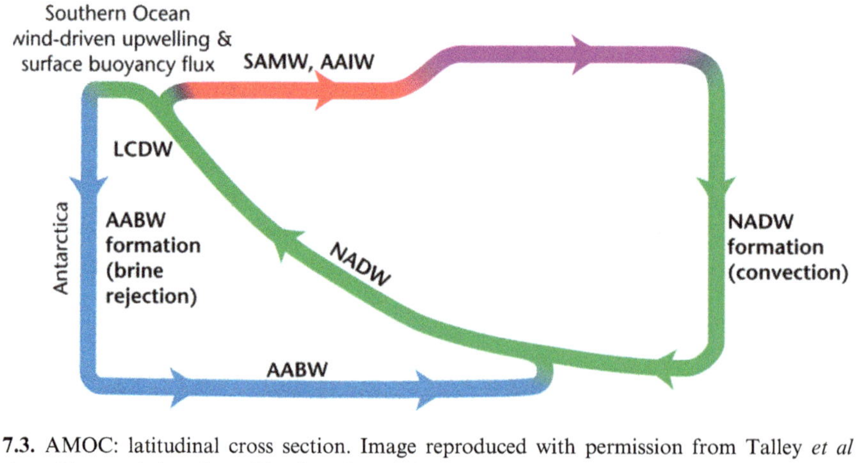

Figure 7.3. AMOC: latitudinal cross section. Image reproduced with permission from Talley *et al* (2011), Copyright, with permission from Elsevier. Copyright © C A Vreugdenhil and B Gayen, Attribution 4.0 International (CC BY 4.0)

zone, on any other supposition than that of cold currents from the poles; and the utility of these currents in tempering the excessive heats of these climates is too evident to require any illustration.' The NADW reaches the ocean depth, activating the AMOC and the mechanism through which it forms and sinks, as has been described by Marshall and Schott (1999). In particular, Rudels and Quadfasel (1991) focused on convection and deepwater formation in the Arctic Ocean-Greenland Sea System. Three phases are identifiable to activate a deep convection (figure 7.4):

1. *Preconditioning linked to cyclonic circulation*: Cyclonic gyres and winter cooling lead to dome-shaped structures with isopycnals sloping up towards the sea surface. This mechanism affects areas in the order of 100 km and can last for months before the meteorological forcing 'breaks' the stratification.
2. *Violent convection*: Once the vertical density stratification is significantly weakened, convection quickly occurs due to further surface cooling: During this phase of convection, small plumes form and sink; vertical velocities up to 10 cm s^{-1} have been measured during convection in location that take the name of 'chimneys' because of their shape. Chimneys connect the ocean surface with depths of up to 1500 m (Wadhams *et al* 2002).
3. *Lateral exchanges*: The lateral advection leads to geostrophic flow: the main mixing becomes lateral through the formation of rim currents.

A fourth phase is sometimes included: *restratification*. Once the convection is over, the restratification process begins, which lasts from weeks to months.

The mechanism described is of fundamental importance to the modulation of the strength of the AMOC: Rahmstorf *et al* (2015), studied the impact of the melting on the Greenland ice sheet on the AMOC slowdown and found a positive correlation between these two processes. The freshening of the surface layers of the Arctic Ocean creates buoyant water, therefore deep convection is inhibited. They also found a strong correlation between the AMOC strength and an AMOC index depending on air temperature (figure 7.5).

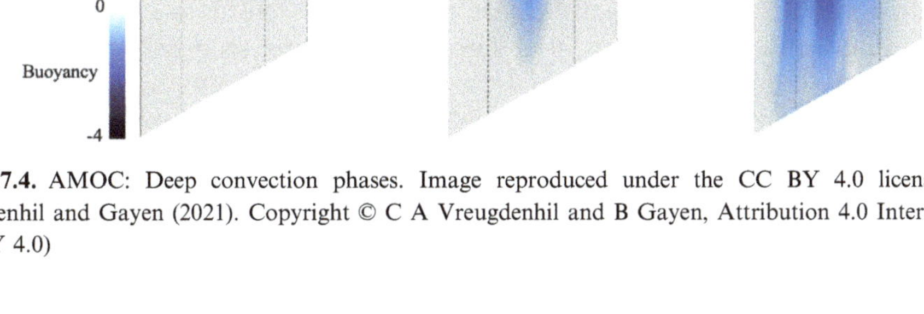

Figure 7.4. AMOC: Deep convection phases. Image reproduced under the CC BY 4.0 license from Vreugdenhil and Gayen (2021). Copyright © C A Vreugdenhil and B Gayen, Attribution 4.0 International (CC BY 4.0)

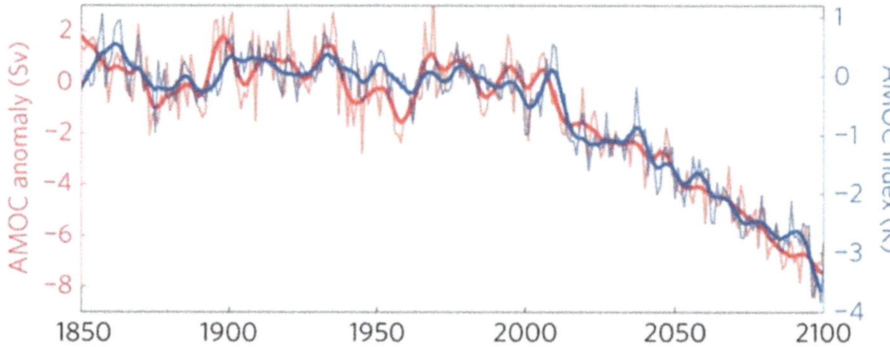

Figure 7.5. AMOC slowdown. Reprinted/adapted with permission from Rahmstorf *et al* (2015). Copyright © (2015).

7.1.4 The AMOC: slowdown or shutdown?

The AMOC is the main large-scale mechanism responsible for the latitudinal balance of heat transport from the equator to the poles. The maintenance of the AMOC is linked to the present state of our climate, since one of its main drivers is the rate of formation of the NADW. In fact, this is one of the critical processes identified by the Intergovernmental Panel on Climate Change that could change completely in its behaviour if a new state of our climate was reached. A complete shutdown of the AMOC would have a strong cooling effect on the climate of Northern Europe (Jackson *et al* 2015), leading to the following:

- Widespread cooling throughout the North Atlantic and Northern Hemisphere in general, with cooling in Europe of several degrees.
- Much greater sea ice coverage in the North Atlantic.
- Less precipitation and evaporation in the Northern Hemisphere mid-latitudes.
- Large changes in precipitation in the tropics with a southwards shift of the Atlantic Intertropical Convergence Zone.
- A strengthening of the North Atlantic storm tracks.

A 2015 study detailed even further the consequences of the AMOC shutdown on the climate of Europe, listing the following particularly significant outcomes:

- Summer precipitation decreases across much of Europe, however there is an increase in precipitation around the Mediterranean. This pattern is associated with a negative summer North Atlantic Oscillation.
- In winter, the storm tracks across the North Atlantic and into Europe are strengthened and penetrate further over land (as seen before). The higher resolution of this model enables us to see very localised regions of increased precipitation likely caused by the increase in winter storms, despite the general signal of decreased winter precipitation across Europe.
- A greater proportion of precipitation falling as snow over all Europe. There is also an increase in the number of months with significant snow cover.

- In both winter and summer the atmospheric circulation response moderates the cooling over central Europe: in winter the westerly winds strengthen, enhancing the winter maritime warming effect, while in summer the westerly winds weaken, which weakens the summer maritime cooling effect. The cooling over Europe is also moderated by reduced cloud cover over land that allows more shortwave radiation to reach the surface, but this is offset by increased albedo from greater coverage of snow and ice.
- River flow and surface water runoff in Europe are significantly reduced because of the reduction in precipitation, however they are also affected by the distribution of precipitation changes and changes to the snowmelt timing.
- Vegetation and crop productivity show strong decreases over Europe in response to the cooling and decrease in available water.

Even though a shutdown of the AMOC would lead to a negative feedback effect over global warming, several studies suggest that a return to a cooler climate might not be enough to reactivate the AMOC (Alkhayuon *et al* 2019). The Special Report on the Oceans, Cryosphere and Climate Change concludes, though, that a complete shutdown of the AMOC is not likely to occur within this century, but the probability that it may happen rises when looking at simulation scenarios of 2100. It is worth underlining that the AMOC plays an important role in the carbon cycle, acting as carbon sink. In particular, a 2016 study suggests that the dissolved organic carbon downward export due to overturning circulation acts as a carbon sink and represents a considerable contribution to CO_2 sequestration: some $\sim$30% of the total North Atlantic CO_2 sink (Fontela *et al* 2016). Many studies show that a slowdown of the AMOC, on the other hand, is already taking place, and it depends strongly on the amount of freshwater input in the Arctic (Jackson and Wood 2018). This would inhibit the activation of the formation of the NADW and deep convection.

A comprehensive review article on the AMOC (Weijer *et al* 2019) is recommended for further reading and to find scope for future work and study topics related to the changes in the AMOC and their consequences to our climate and health.

7.2 Ocean ventilation

Deepwater formation is linked to another important phenomenon that occurs mainly at high latitudes: deep ocean ventilation. Young waters that have been in touch with the atmosphere sink, activating a downward transport of surface waters that propagates into the ocean depths the properties (e.g. temperature, salinity and concentrations of dissolved gases) imprinted by the atmosphere on the ocean surface in the areas of ventilation and then towards the low latitudes (Naveira Garabato *et al* 2017). When water masses form and then sink, they retain information about the atmosphere from previous ocean interactions. One of the main characteristics of young water recently in contact with the atmosphere is its high quantity of oxygen (Naveira Garabato *et al* 2017). In fact, the characteristics of a body of water in terms of their content in salt, oxygen and nutrients can be used as a 'signature' that helps oceanographers to retrieve the origin of the water masses and often the location of their formation. It has been estimated that up to two-thirds of the oceans' waters has gone through processes of ventilation at high latitudes (Khatiwala *et al* 2012), making these areas of fundamental importance for the maintenance of our climatic equilibrium.

7.3 High-latitude meteorological indexes

7.3.1 The North Atlantic Oscillation

The strength of the AMOC is dependent on the strength of the North Atlantic Oscillation (NAO) index. Its drivers, as already explained, are linked to dense-water production at high latitudes. The preconditioning situation that leads to the formation of the NADW is dependent on meteorological factors, being linked to mid-scale cyclonic gyres that induce loss of stratification in the ocean. This pattern is more frequent in winter, and it is linked to a positive NAO phase (Smethie *et al* 2000). The National Centre for Environmental information defines the NAO index as '*the surface sea-level pressure difference between the subtropical (Azores) high and the subpolar low*'. The strength of the subpolar low is determined by the behaviour of the polar jet stream. Jet streams can be described as channels of strong winds moving eastwards and situated in the upper levels of the atmosphere. They move across bands that divide hot and cold air: technically they behave as thermal winds. The strength of the subpolar low influences the latitude band within which the jet stream is moving, i.e. in the Northern Hemisphere, weaker than normal subpolar lows lead to large meanders and a large latitudinal variation in the jet stream (figure 7.6). This is turns lead to a greater/smaller pressure and temperature gradient between Iceland and the Azores, meaning, respectively, a positive/negative phase of the NAO.

Positive/negative phases of the NAO index are associated with large-scale processes with long-term consequences on Europe and North America because they influence the strength of the AMOC and the location and intensity of the jet stream. On a shorter timescale, the phases of the NAO also influence the weather in

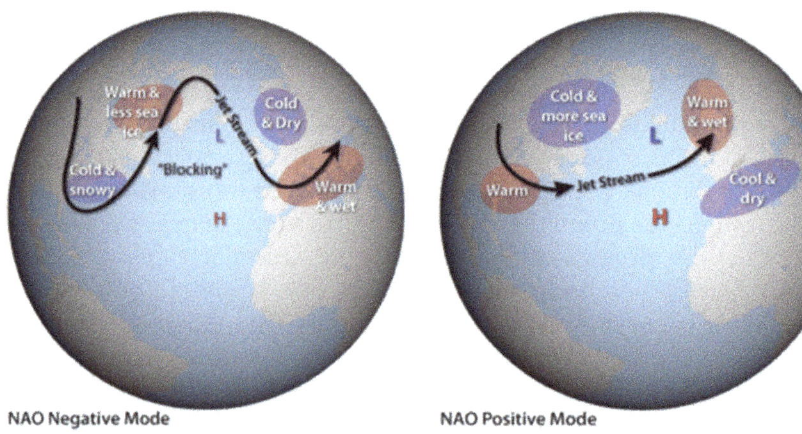

Figure 7.6. Jet stream and NAO. Schematic adapted from AIRMAP by Ned Gardiner and David Herring, NOAA.

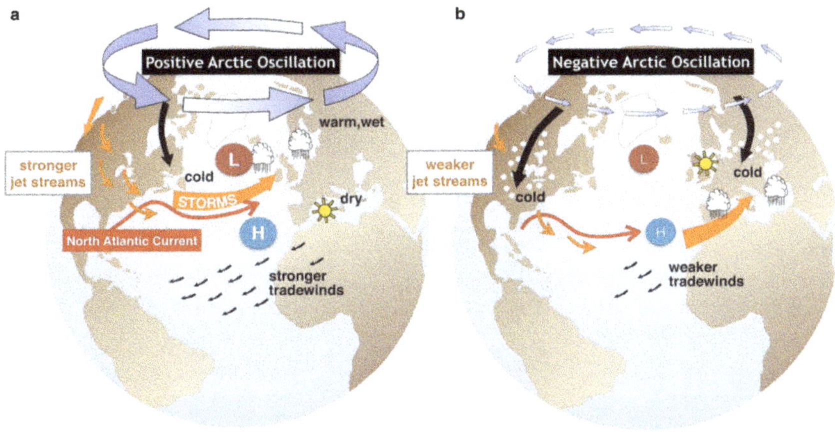

Figure 7.7. NAO phases: effects on the Northen Hemisphere climate.

Northern Europe, and since jet streams follow the boundaries between hot and cold air, they become stronger when the temperature gradient is at its maximum, thus mainly in winter.

In winter, a positive NAO index phase is associated with warmer-than-average temperatures across Northern Europe and Northern Asia as well as southeast North America, and lower-than-normal temperatures in North Africa, northeast Canada and southern Greenland (7.7). It also leads to stormy weather across Northern Europe and drier conditions in Greenland and Europe.

During negative NAO phase the reverse weather is observed: temperatures are on average cooler and fewer storms reach Northern Europe. Southern Europe will experience rainy conditions, while the eastern coast of the United States will have drier weather. More rain will be observed in Canada and the western coasts of Greenland.

In general, a positive NAO is associated with a more stable and stronger polar vortex, while a negative NAO is associated with a 'wavier', less stable polar vortex that presents deeper troughs and leads to unstable weather conditions.

7.3.2 The Southern Annular Mode index

The Southern Hemisphere's climate is characterised by three large-scale drivers: El Niño, La Niña and the Indian dipole. As for the Northern Hemisphere, though, the position of the polar jet stream influences the climate in Australia: this is known as the Southern Annular Mode (SAM). The SAM, similar to the NAO index, is defined as the difference of zonal mean sea-level pressure between 40°S and 65°S. In winter, the southward shift of the polar front corresponds to a positive SAM index and leads to drier winters in southern Australia, while in the north rainier weather can be expected. During negative SAM phases, the polar front brings rainy weather along the south of Australia. In the summer, when stronger high pressure exists over Australia, the consequences of positive/negative SAM index are inverted compared to the winter ones. The Australian Bureau of Meteorology describes well the effect

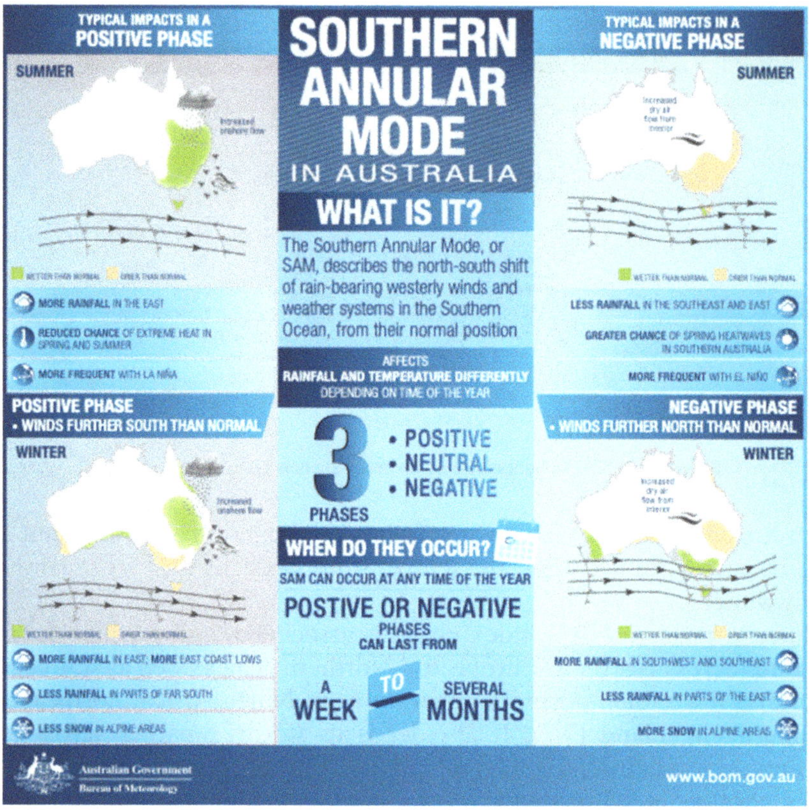

Figure 7.8. Southern Annular Mode phases: effects on the Australian climate. This material is sourced from the Australian Bureau of Meteorology.

that the three possible SAM phases (positive, negative and neutral), can have on the weather in Australia depending on the season (figure 7.8). A time series of the SAM index has been retrieved using weather station data and shows an increase of the frequency of positive phases of the SAM index in the past three decades (Trenberth *et al* 2007), shown in figure 7.9.

7.4 The poles and the changing climate

7.4.1 The Arctic

The Arctic can be described as the area of the Earth north of 60°N. It includes the Arctic ocean and parts of Canada and Siberia. Climatically, the Arctic is often defined as the area north of the 10 °C July isotherm, i.e. north of the line or region which has a mean July temperature of 10 °C (AMAP 1998).

The Arctic is surrounded by land and has two connections with the global ocean: one with the Pacific Ocean through the Bering Strait and one with the Atlantic Ocean through the Fram Strait. Through the Canadian Archipelago, the Arctic Ocean is connected to the Labrador Sea. The Arctic Ocean is partially covered by

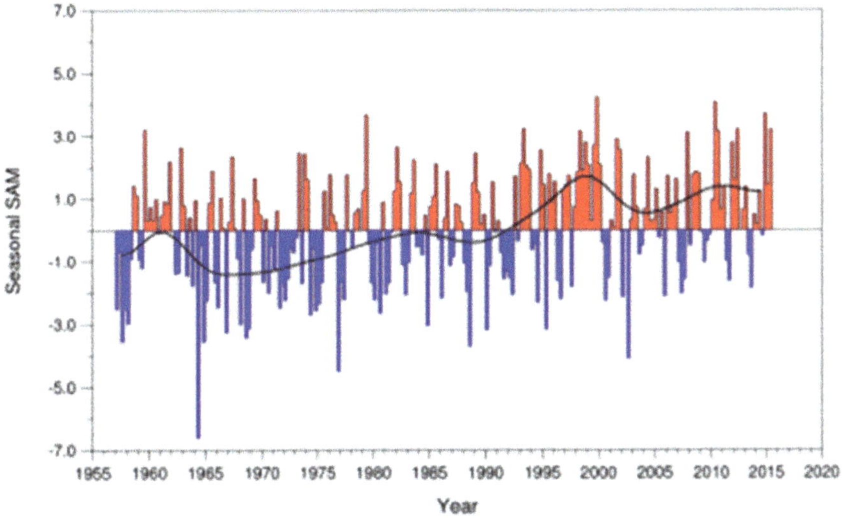

Figure 7.9. Seasonal values of the observation-based SAM index. The smooth black curve shows decadal variations. This figure is updated from Figure 3.32 in Trenberth *et al* (2007) (contributed by G Marshall 2005).

permanent sea ice pack. The Arctic is probably the region on Earth which has experienced the most dramatic effects of global warming: sea ice decline has seen many locations transition from a multiyear to a seasonal ice cover.

In the 21st century the Northwest Passage connecting the Atlantic to the Pacific along the Canadian Arctic has been ice-free in several seasons when sea ice reached its minimum. Given the difficult and costly logistics related to this still remote environment, this is not yet seen as a viable alternative to the current commercial shipping routes (figure 7.10). Similarly, part of the Northeast Passage has remained ice-free throughout the year (Buixadé Farré *et al* 2014), but also in this case several concerns were raised, especially from an environmental standpoint. In September 2012, Inuit Circumpolar Conference Chair Jimmy Stotts was reported as saying there is concern that increased shipping could adversely affect indigenous hunting of marine mammals. The potential environmental effects of increased shipping activities in the Northwest Passage and the Beaufort Sea include higher risk of oil or waste spills, changes in ice conditions due to leads caused by ship wakes, and impacts on wildlife and marine species (Stewart *et al* 2005).

7.4.2 Arctic sea ice and climate

The seasonal sea ice cycle sees a maximum extent of sea ice cover in March and its minimum in September. The September minimum has been identified as a proxy of sea ice decline caused by air temperature rise. Maintenance of the sea ice system results from a balance of atmosphere–ice-ocean thermodynamic and dynamic processes, and the causes of the observed sea ice reduction are complex (Perovich *et al* 2008). Larger areas of open water observed in the summer increase the heat storage in the ocean, leading to increased water temperatures and resulting in

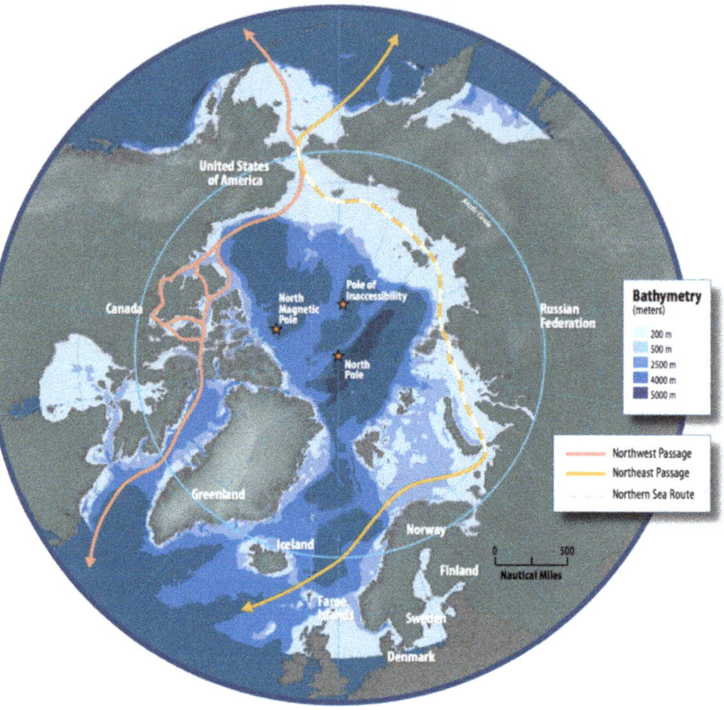

Figure 7.10. The Arctic region showing the Northeast Passage, the Northern Sea Route within it, and the Northwest Passage. Public domain image reproduced from Ellis and Brigham (2009). Copyright Susie Harder.

additional bottom melt of sea ice (Perovich *et al* 2008, Perovich *et al* 2009, Tsamados *et al* 2015) and a consequent delay in winter sea ice formation.

In its report card from 2021, the NOAA highlighted the following points on the most recent sea ice observations (Meier *et al* 2021):

- Winter maximum (March) and summer minimum (September) 2021 sea ice extents were less extreme compared to the last couple of years, but the 15 lowest minimum extents have all occurred in the last 15 years.
- An early onset of sea ice melt and retreat occurred in the Laptev Sea, leading to record low extent in the region during May and June, while winter advection of thick, multiyear ice into the Beaufort and Chukchi seas resulted in a late and limited spring retreat in that region.
- The amount of multiyear sea ice, based on available data since 1985, reached its second lowest level by the end of summer 2021, sea ice thickness was lower than recent years, and volume was at record low (since at least 2010) in April 2021.

The September Arctic sea ice is now declining at a rate of 13% per decade, relative to the 1981 to 2010 average. Figure 7.11 shows the annual Arctic sea ice minimum each March and September since 1979, derived from satellite observations (Meier *et al* 2021). Arctic sea ice thickness has also declined, as shown by data retrieved from measurements collected by several satellite missions (Figure 7.12).

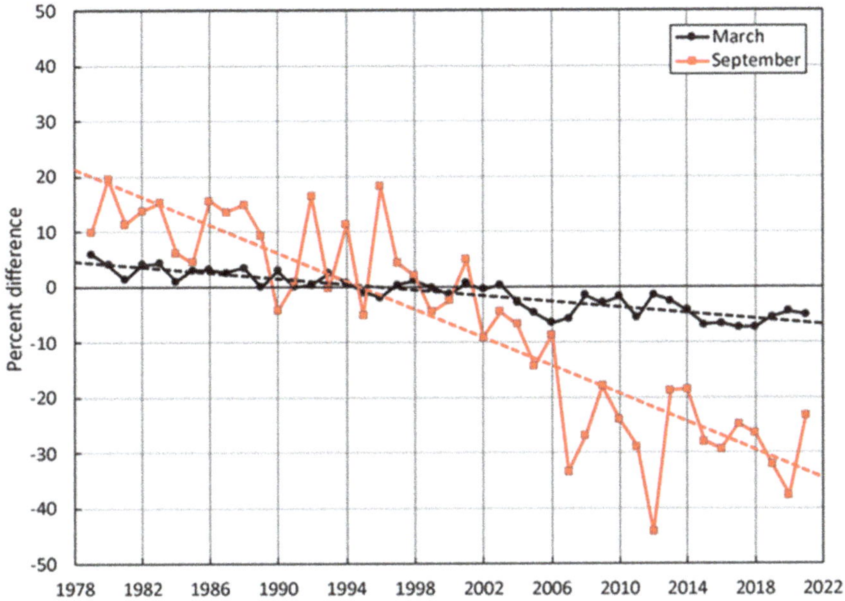

Figure 7.11. Monthly sea ice extent anomalies (solid lines) and linear trend lines (dashed lines) for March (black) and September (red) 1979 to 2021. The anomalies are relative to the 1981 to 2010 average for each month. Figure freely reproduced courtesy of NOAA from Meier *et al* (2021).

A striking way to show the sea ice volume decline is the so-called 'sea ice death spiral', reconstructing a time series from 2014 to present using data from the PIOMAS Arctic Sea Ice Volume Reanalysis (figure 7.13).

The thermodynamic processes that have led to such a steep decline in sea ice volume are related to the enhanced air temperature rise observed in the Arctic. The Arctic has in fact undergone larger temperature anomalies compared to the Earth average; this phenomenon is defined as polar amplification. A measure of this process is the polar amplification factor, defined as the ratio between the temperature gradient in the Arctic compared to the average Earth temperature gradient in a defined amount of time. Figure 7.14 shows how the temperature anomaly in the Arctic can be between two and three times the anomaly observed in other regions on Earth.

The reason for the occurrence of this phenomenon is mainly identified in the positive ice–albedo feedback mechanism for which the sea ice decline is responsible. For a given radiation imbalance at high latitudes, sea ice melts, forming melt ponds with lower albedo, that in turn lead to further melting. Furthermore, when sea ice melts, larger areas of the ocean with low albedo are exposed to the atmosphere, absorbing more solar radiation, warming up and leading to possible melting of the ice bottom.

The ice–albedo feedback mechanism is also valid in the opposite direction: a cooling of the Earth temperature would lead to an increased sea ice formation, with consequent rise of the average albedo, leading to further cooling. An interesting

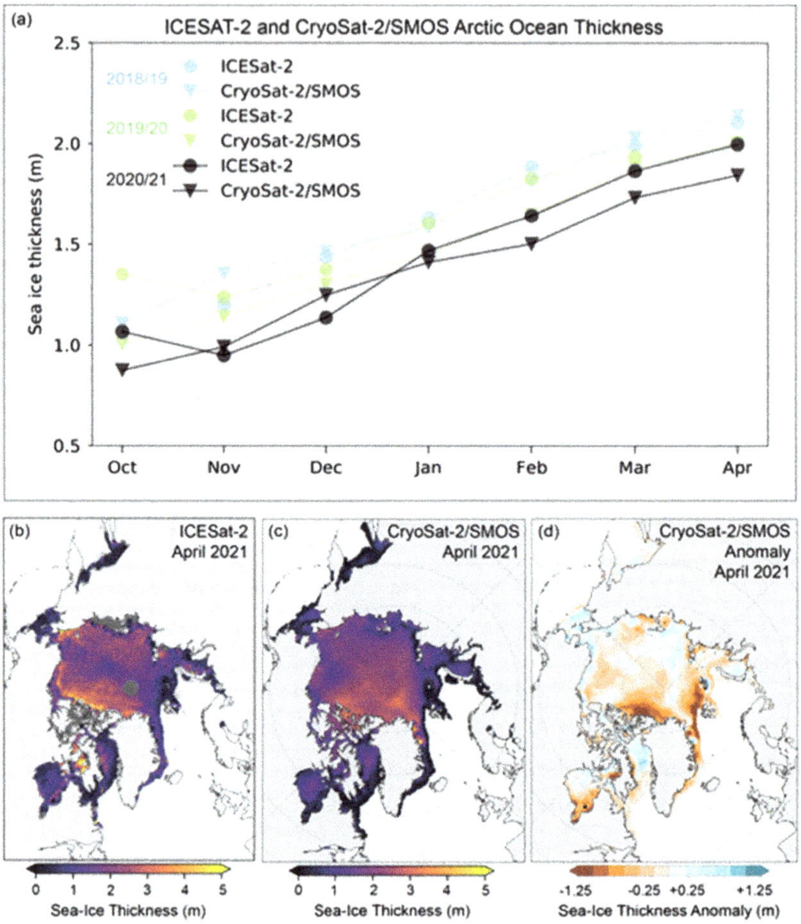

Figure 7.12. October through April monthly average sea ice thickness, calculated over an Inner Arctic Ocean Domain from ICESat-2 (circles) and CryoSat-2/SMOS (triangles) for 2018–19 (blue), 2019–20 (green) and 2020–21 (dark grey). Average April 2021 sea ice thickness maps from (b) ICESat-2 (dark grey areas have no data) and (c) CryoSat-2/SMOS; (d) CryoSat-2/SMOS thickness anomaly (relative to the 2010–20 average). Note that ICESat-2 thickness estimates outside the Arctic Ocean Domain are not as reliable due to uncertainties in snow cover. Figure freely reproduced courtesy of NOAA from Meier *et al* (2021).

theory is the one that explains the 'snowball Earth' through this mechanism (Kirschvink 1992). This hypothesis has been considered to explain the presence of glacial sediments at low latitudes, through a near total Earth glaciation, which would have taken place before 650 million years ago during the Cryogenian period.

The scientific debate over this hypothesis is still open, since current general circulation models are not able to recreate the condition for simulations that would allow sea ice formation at tropical latitudes (Poulsen *et al* 2001). Furthermore, there is no definitive answer that explains how the Earth exited its total ice-covered status. One hypothesis postulates that volcanic activity together with bacterial respiration

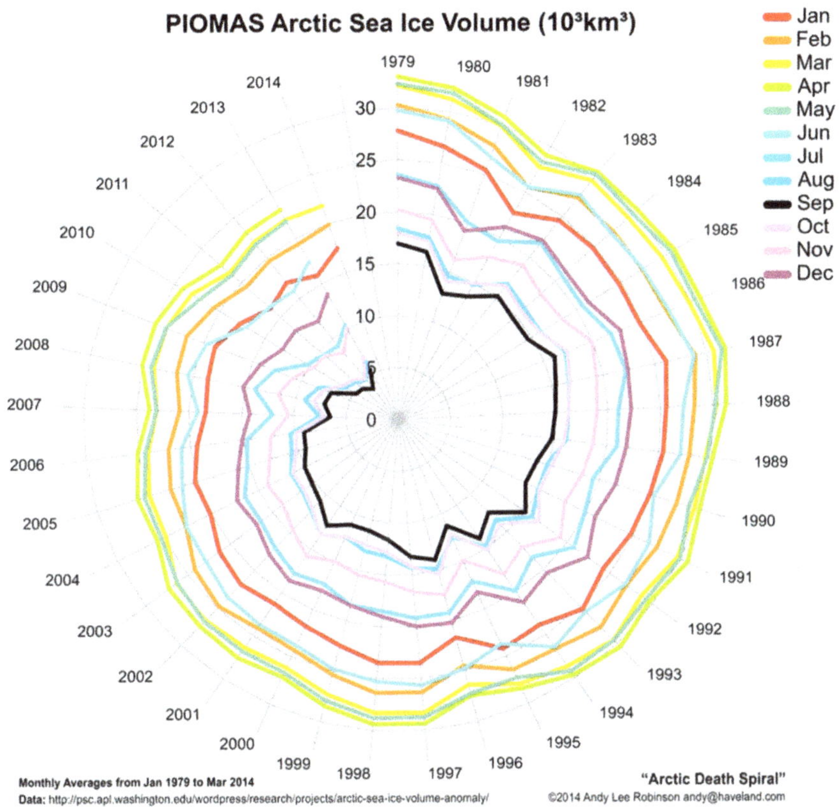

Figure 7.13. Sea ice death spiral: data from the PIOMAS Arctic Sea Ice Volume Reanalysis. Image created by Andy Lee Robinson and reproduced under a CC BY-SA 4.0 licence. (CC BY 4.0)

would have introduced large amounts of greenhouse gases in the atmosphere, thereby inducing a warming process (Crowley *et al* 2001).

7.4.3 Antarctica and the Southern Ocean

The Antarctic continent is roughly centred at the South Pole and covers an area of just over 14 million km^2. Contrary to the Arctic, the formation of sea ice around Antarctica is not constrained by land presence. The Antarctic continent is surrounded by the Southern Ocean, which the International Hydrographic Organization only recognised as an ocean in 2021. It comprises the southernmost waters of the World Ocean, generally taken to be south of 60°S latitude and encircling Antarctica. The Antarctic continent is almost completely covered by ice, which can reach a thickness of almost 5 km with an average of over 2 km. Complete melt of the Antarctic ice cover would potentially lead to a sea water level rise of 60 m.

The mechanisms of the maintenance of sea ice extent in Antarctica have been much studied because for years they represented a paradox, because despite warming temperatures, the maximum extent of sea ice had been rather constant or slightly increasing until 2015. Ever since then, subsequent minima in February

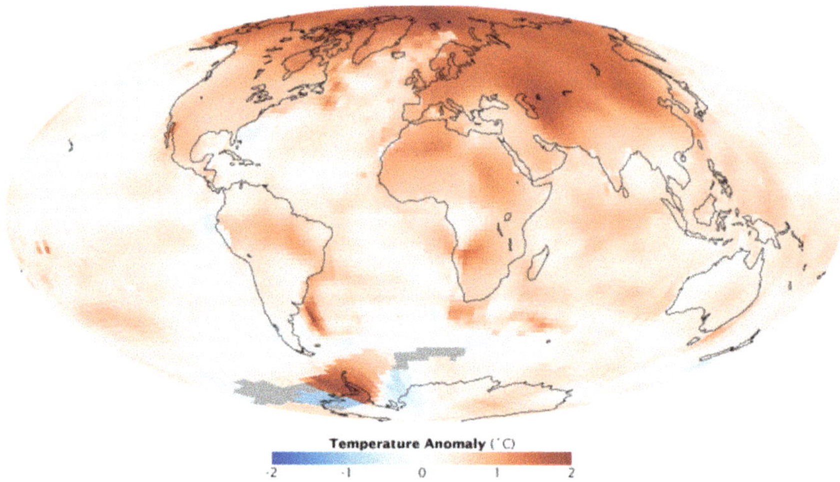

Figure 7.14. NASA GISS temperature trend 2000–09, showing strong Arctic amplification. NASA image by Robert Simmon, based on GISS surface temperature analysis data including ship and buoy data from the Hadley Centre. Caption by Adam Voiland and Holli Riebeek.

have been observed, with a lowest ice cover of 2.17 million km^2 recorded in February 2022 (figure 7.15).

The behaviour of the sea ice extent in Antarctica has been linked to the local wind pattern (Holland and Kwok 2012, Holland *et al* 2017) and to the SAM index. The minimum sea ice extent observed in 2017 has been explained by a correlation with the SAM index (Doddridge and Marshall 2017). Several review studies, though, have highlighted the complexity of the ocean–sea ice–atmosphere system in the Southern Hemisphere and exposed the need to further study the dynamics of the sea ice in the Southern Ocean (Hobbs *et al* 2016, Jones *et al* 2016).

7.4.4 Description of the Antarctic Circumpolar Current

The largest current on Earth, the Antarctic Circumpolar Current (ACC), travels clockwise around the Antarctic continent. This makes it the only current that circles the entire globe, as shown in figure 7.16. Even though it is not fast, it transports the most mass: a mean ACC transport of 100–150 Sverdrups (1 Sv = 106 m^3 s^{-1}) has been retrieved by observations gathered south of Tasmania and New Zealand (Knauss 1978). The only geographical constraint that funnels the ACC is the Drake Passage, south of Cape Horn. From an oceanographic point of view, the sub-Antarctic front (SAF), is considered the northernmost boundary of the ACC and it is characterised by sea surface temperatures near 7 °C–9 °C and a salinity minimum of 33.0–34.0 psu.

The Circumpolar Deep Water (CDW), a mix of intermediate and bottom water that blends with the Pacific and Indian waters, sits at a depth of about 500 m and plays an important role in the ACC: being relatively salty and warm, it contributes to the melting of the bottom of ice shelves (Dinniman *et al* 2011). The ACC flows far from the continent in the areas of the Ross Ice Shelf and the

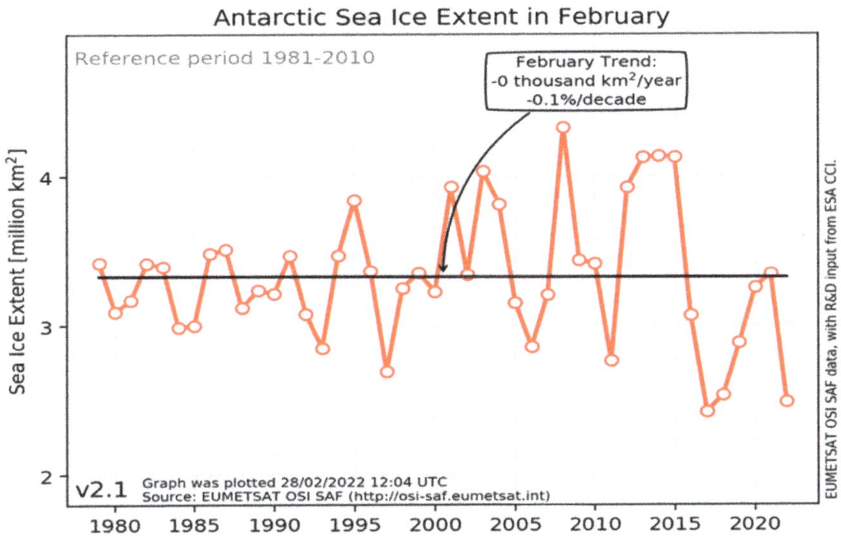

Figure 7.15. Antarctic sea ice extent, retrieved from OSI-SAF—EUMETSAT. Copyright © 2021 EUMETSAT.

Weddell Ice Shelf where two gyres[1] keep the ACC at distance. In other areas, such as the Bellinghausen and Amundsen seas, its flow interacts more closely with the continent.

7.4.5 Deepwater formation

The Antarctic Bottom Water (AABW) is the densest water on Earth and fills the depth of the world oceans, contributing to activation of convection in the Southern Ocean, similarly to the NADW in the Arctic. Its formation closes the circle of the AMOC, as already shown in figure 7.3. It mainly forms along the coasts of the Weddell Sea and the Adélie Coast in the Ross Sea, strongly linked to the presence of a high sea ice production rate in polynya areas. Polynyas occur due to the strong mechanical action of the katabatic winds over the ocean waters, therefore exposing them to low air temperatures. This in turns leads to an enhancement of ice production rates and salt release in the column of water underneath. In Terra Nova Bay, high-salinity shelf water is formed through this mechanism: this is the densest water of the Ross Sea. The other mechanism of formation of AABW happens under the ice shelves: when a body of water flows under an ice shelf it melts some of the ice entraining freshwater at freezing temperature, thus forming the so-called ice shelf waters, characterised by temperatures close to freezing point.

[1] A gyre is any large system of rotating ocean currents.

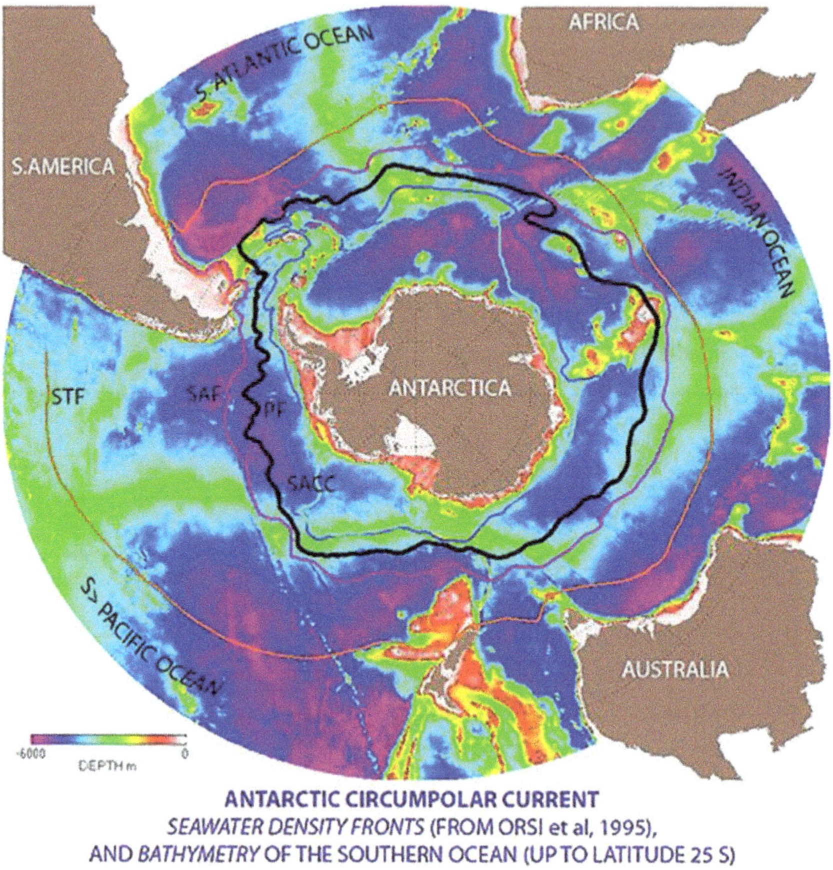

Figure 7.16. ACC image from the Grace Mission. SACC = southern Antarctic Circumpolar Current front; PF = polar front; SAF = sub-Antarctic front; STF = subtropical front. Courtesy of NASA/JPL-Caltech.

7.4.6 The ozone hole

Ozone (O_3) is present in the Earth's atmosphere at various altitudes. In the troposphere ozone is the product of pollution and shares the same characteristics as greenhouse gases, absorbing part of the Earth's infrared radiation (figure 7.17). Ozone is naturally found in the lower layer of the stratosphere and acts as a major filter of the fraction of solar radiation harmful to humans, as rays of UV-C, characterised by wavelengths of 280–100 nm.

In 1985 Jonathan Shanklin, a researcher at the British Antarctic Survey, in Cambridge (UK), observed a 'hole' in the stratospheric ozone layer over the Antarctic, raising an alarm that reverberated through the scientific community and had huge repercussions on the public. CFCs, among other compounds, release elevated amounts of halogens when they reach the stratosphere through mixing, which act as catalysts for a chemical reaction that turns ozone into oxygen. Shanklin's findings led to the Montreal Protocol treaty that implemented actions to reduce the use of CFCs from 1996. The protocol foresees a complete phase-out of CFCs production by 2030.

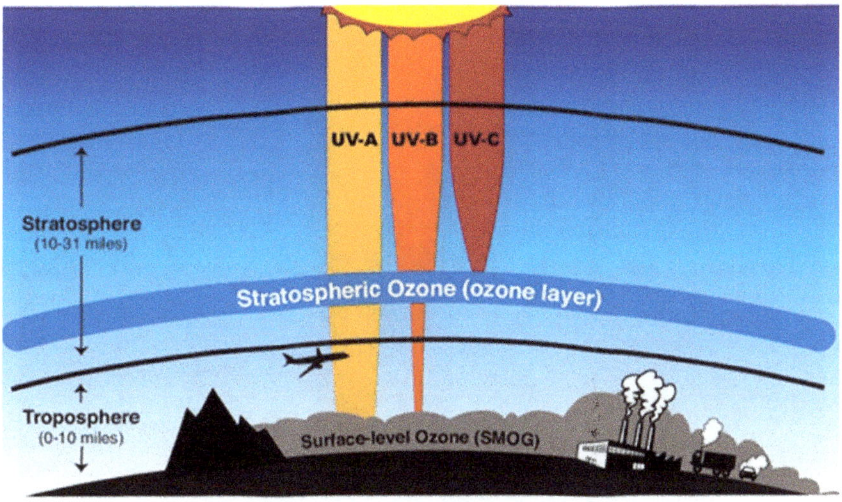

Figure 7.17. Stratospheric ozone layer schematic. Image courtesy of NASA.

Ever since its creation in 2015, the Copernicus Atmosphere Monitoring Service (CAMS) provides consistent and quality-controlled information related to air pollution and health, solar energy, greenhouse gases and climate forcing, everywhere in the world. CAMS combines measurements from satellite instruments and *in situ* sensors with its numerical models to provide quality-assured information about the state of the ozone layer. The ozone layer is slowly recovering, as expected, compared to the '80s: a minimum was observed in 2019, followed by an anomalous 2020 and 2021, when the ozone hole has been larger and deeper than in the previous years, due to meteorological forcing over the Antarctic continent that enhanced ozone depletion.

7.5 Polar lights: Aurora Borealis and Australis

A beautiful display of lights called the Aurora Borealis or Australis, respectively, in the Northern or Southern hemispheres appears at high latitudes during the polar night. Polar lights are due to interactions between the Earth's magnetosphere and the solar wind (figure 7.18). Their location is mainly restricted to the high latitudes because of the shape of the Earth's magnetic field that generally shields the Earth from the ionised particles incoming from the Sun; at the poles, the Earth's magnetic field follows lines that 'enter' the poles, and in these locations some of the solar particles enter the magnetosphere, interacting with gases and releasing energy in the form of light (figure 7.19).

To the public, they are visible testimony of great beauty and are famous as a tourist attraction, being one of the seven natural wonders in the world. The colour of the auroras depends on the gases that are present in the atmosphere, so that when blue is observed it means that the solar wind is encountering nitrogen atoms, while when the auroras display a yellow-green colour, it means that the most present gas is oxygen. As can be seen in figures 1.1 and 7.19(c), auroras are sometime visible by satellite.

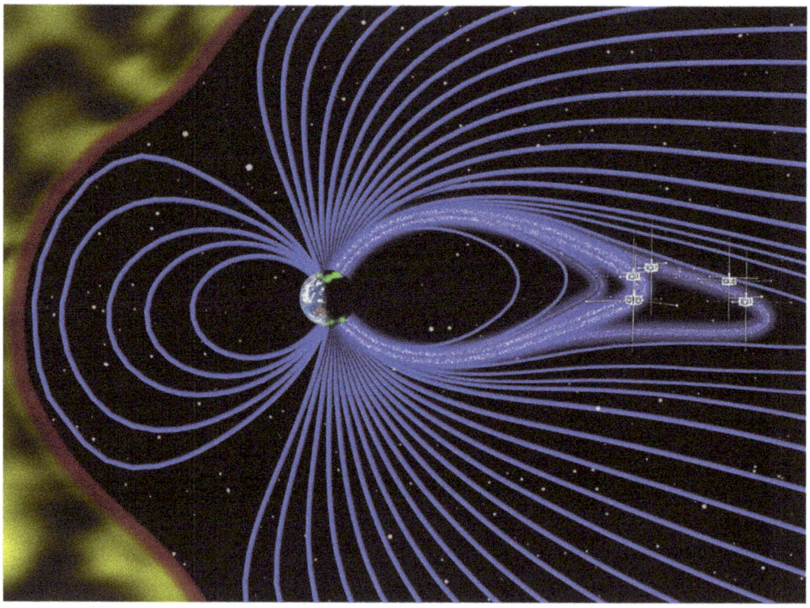

Figure 7.18. The Earth's magnetosphere and its interaction with the solar wind. Image credit: Emmanuel Masongsong/UCLA EPSS/NASA.

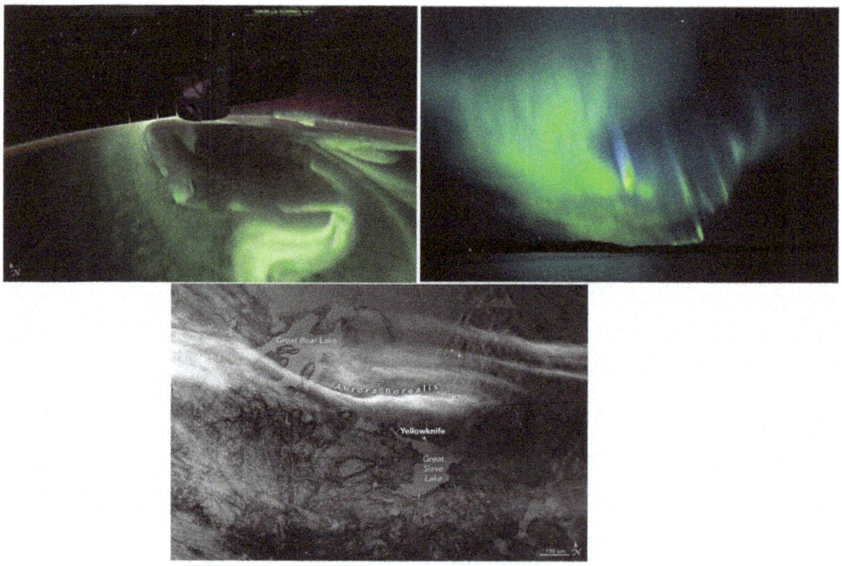

Figure 7.19. Polar lights: (a) Aurora australis seen from the ISS, 2017. Green auroras indicate collisions with oxygen; (b) Northern Lights with very rare blue light emitted by nitrogen; and (c) Aurora image acquired on February 4, 2020, by the Visible Infrared Imaging Radiometer Suite (VIIRS) on the NOAA-NASA Suomi NPP satellite

7.6 Ecology and sea ice

The Arctic is home to an indigenous population that reaches four million people and is home to numerous species of mammals, birds and fish that are dependent on sea ice to thrive. In the Antarctic there is no dwelling human population, but a relatively high biodiversity is observed, especially in the ocean. Biodiversity shows a latitudinal gradient increasing towards the equator, where it reaches its maximum (figure 7.20).

Animal species living at the poles have developed highly specialised adaptation strategies to survive the extreme climate; but this also means that they might find it more difficult to adapt to changes in their ecosystem, especially if these changes are as rapid as the ones observed in current days.

Polar bears, for example, depend on sea ice directly and indirectly. They use sea ice as a platform to hunt and raise their cubs, therefore ice fragmentation will cause disruption to their feeding and breeding habits; and they rely upon sea ice indirectly because their diet is dependent on the lower levels of the food web (from phytoplankton to small fish), as determined by a study that confirmed that polar bears feed more on sympagic (i.e. sea-ice-associated) rather than pelagic sources of food (Brown *et al* 2018). Thus, current and predicted reductions in sea ice occurrence and persistence are likely to have major impacts on their survival (Stirling and Derocher 1993, Laidre *et al* 2022). Unfortunately, photographs of skinny and stranded polar bears are already often seen in the media, such as the one shown in figure 7.21.

In the Antarctic biodiversity on land is rather limited and mainly observed in ice-free areas. In the ocean, though, phytoplankton blooms are abundant and frequent, and are linked to the ocean dynamics around the continent. The CDW plays an important role in supporting Antarctic ecosystems: upwelling of nutrients-rich

Figure 7.20. Latitudinal diversity gradient. Graphic courtesy of Julie McMahon.

Figure 7.21. A starving polar bear. Copyright © Andreas Weith, Attribution-ShareAlike 4.0 International (CC BY-SA 4.0).

Figure 7.22. Phytoplankton bloom in Terra Nova Bay. Copyright © European Union and the Copernicus component of the EU Space programme.

waters occur in the Antarctic Convergence Zone and along the Antarctic coasts, creating the conditions for large phytoplankton blooms that support the Antarctic food web (figure 7.22).

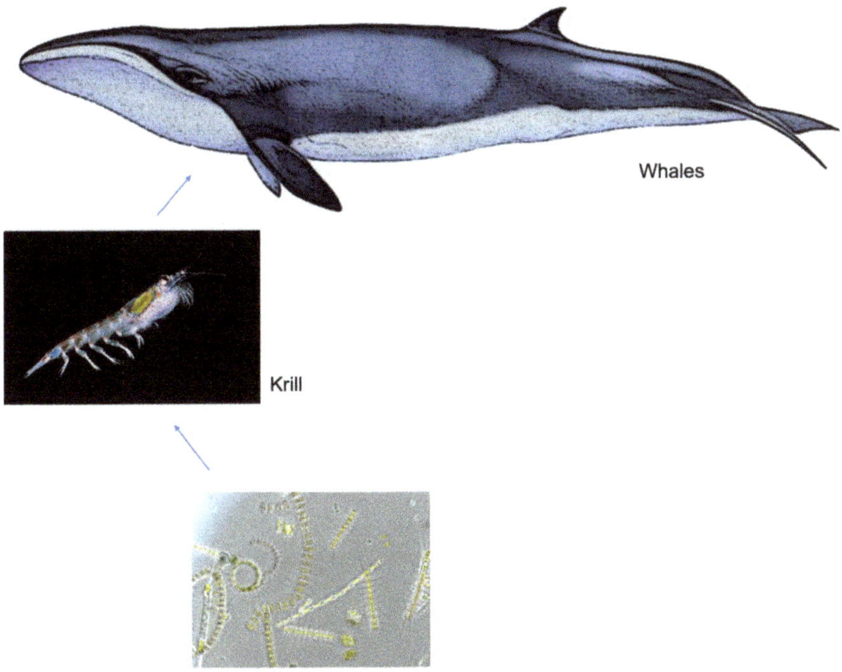

Figure 7.23. A simple Antarctic food chain.

One of the most efficient food chains is found in the Antarctic: short food chains mean the least amount of energy loss, even though the maintenance of this food chain is strictly dependent on the availability of krill (figure 7.23).

The expected change in sea ice conditions in the Antarctic may influence the whole dynamic of the Antarctic food chain (Post *et al* 2013), since sea ice provides a habitat for microscopic marine algae, released in the summer when the ice melts and which are fed upon by krill (Rossi *et al* 2019). Rapid changes in the sea ice conditions would therefore lead to damaging consequences to a unique and fragile ecosystem like the Antarctic continent.

References

Alkhayuon H *et al* 2019 Basin bifurcations, oscillatory instability and rate-induced thresholds for Atlantic meridional overturning circulation in a global oceanic box model *Proc. R. Soc. A: Math. Phys. Eng. Sci.* **475** 20190051

AMAP *et al* 1998 Arctic pollution issues *AMAP Assessment Report* Arctic Monitoring and Assessment Programme (AMAP)

Brown T A *et al* 2018 High contributions of sea ice derived carbon in polar bear (*Ursus maritimus*) tissue *PLoS One* **13** e0191631

Buixadé Farré A *et al* 2014 Commercial Arctic shipping through the Northeast Passage: routes, resources, governance, technology, and infrastructure *Polar Geogr.* **37** 298–328

Crowley T J, Hyde W H and Peltier W R 2001 CO_2 levels required for deglaciation of a 'near-snowball' Earth *Geophys. Res. Lett.* **28** 283–6
Dinniman M S, Klinck J M and Smith W O 2011 A model study of Circumpolar Deep Water on the West Antarctic Peninsula and Ross Sea continental shelves *Deep Sea Res. Part II: Top. Stud. Oceanogr.* **58** 1508–23
Doddridge E W and Marshall J 2017 Modulation of the seasonal cycle of Antarctic Sea Ice extent related to the southern annular mode *Geophys. Res. Lett.* **44** 9761–8
Ellis B and Brigham L 2009 Arctic Marine Shipping Assessment 2009 *Report* Arctic Council
Fontela M *et al* 2016 Dissolved organic carbon in the North Atlantic meridional overturning circulation *Sci. Rep.* **6** 26931
Hobbs W R *et al* 2016 A review of recent changes in Southern Ocean sea ice, their drivers and forcings *Global and Planetary Change* **143** 228–50
Holland M M, Landrum L, Kostov Y and Marshall J 2017 Sensitivity of Antarctic sea ice to the Southern Annular Mode in coupled climate models *Clim. Dyn.* **49** 1813–31
Holland P R and Kwok R 2012 Wind-driven trends in Antarctic sea-ice drift *Nat. Geosci.* **5** 872–5
Jackson L C *et al* 2015 Global and European climate impacts of a slowdown of the AMOC in a high resolution GCM *Clim. Dyn.* **45** 3299–316
Jackson L C and Wood R A 2018 Timescales of AMOC decline in response to fresh water forcing *Clim. Dyn.* **51** 1333–50
Jones J M *et al* 2016 Assessing recent trends in high-latitude Southern Hemisphere surface climate *Nat. Clim. Change* **6** 917–26
Khatiwala S, Primeau F and Holzer M 2012 Ventilation of the deep ocean constrained with tracer observations and implications for radiocarbon estimates of ideal mean age *Earth Planet. Sci. Lett.* **325-326** 116–25
Kirschvink J L 1992 *Late Proterozoic Low-Latitude Global Glaciation: The Snowball Earth* (Cambridge: Cambridge University Press)
Knauss J A 1978 *Introduction to Physical Oceanography* (Englewood Cliffs, NJ: Prentice-Hall)
Laidre K L *et al* 2022 Glacial ice supports a distinct and undocumented polar bear subpopulation persisting in late 21st-century sea-ice conditions *Science* **376** 1333–8
Marshall J and Schott F 1999 Open-ocean convection: observations, theory, and models *Rev. Geophys.* **37** 1–64
Meier W N *et al* 2021 Sea ice *NOAA Technical Report* OAR ARC; 21-05 National Oceanic and Atmospheric Administration
Naveira Garabato A *et al* 2017 High-latitude ocean ventilation and its role in Earth's climate transitions *Phil. Trans. R. Soc.* A **375** 20160324
Perovich D K *et al* 2009 Loss of sea ice in the Arctic *Annu. Rev. Mar. Sci* **1** 417–41
Perovich D K *et al* 2008 Sunlight, water, and ice: extreme Arctic sea ice melt during the summer of 2007 *Geophys. Res. Lett.* **35** L11501
Post E *et al* 2013 Ecological consequences of sea-ice decline *Science* **341** 519–24
Poulsen C J, Pierrehumbert R T and Jacob R L 2001 Impact of ocean dynamics on the simulation of the neoproterozoic 'snowball Earth *Geophys. Res. Lett.* **28** 1575–8
Rahmstorf S *et al* 2015 Exceptional twentieth-century slowdown in Atlantic Ocean overturning circulation *Nat. Clim. Change* **5** 475–80
Rossi L *et al* 2019 Antarctic food web architecture under varying dynamics of sea ice cover *Sci. Rep.* **9** 12454

Rudels B and Quadfasel D 1991 Convection and deep water formation in the Arctic Ocean-Greenland Sea System *J. Mar. Syst.* **2** 435–50

Smethie W M Jr *et al* 2000 Tracing the flow of North Atlantic Deep Water using chlorofluorocarbons *J. Geophys. Res. Oceans* **105** 14297–323

Stewart E J, Draper D and Johnston M E 2005 A review of tourism research in the polar regions *Arctic* **58** 383–94

Stirling I and Derocher A E 1993 Possible impacts of climatic warming on polar bears *Arctic* **46** 240–5

Talley L D *et al* 2011 Global circulation and water properties: supplementary materials *Descriptive Physical Oceanography* eds L D Talley *et al* 6th edn (Boston, MA: Academic) ch S14 pp 1–9

Thomas M D *et al* 2021 Oceanic pathways of an active pacific meridional overturning circulation (PMOC) *Geophys. Res. Lett.* **48** e2020GL091935

Trenberth K E *et al* 2007 Observations: surface and atmospheric climate change *Climate Change 2007: The Physical Science Basis. Contribution of Working Group 1 to the 4th Assessment Report of the Intergovernmental Panel on Climate Change* (Cambridge: Cambridge University Press)

Tsamados M *et al* 2015 Processes controlling surface, bottom and lateral melt of Arctic sea ice in a state of the art sea ice model *Phil. Trans. R. Soc.* A **373** 20140167

Vreugdenhil C A and Gayen B 2021 Ocean convection *Fluids* **6** 360

Wadhams P *et al* 2002 A deep convective chimney in the winter Greenland sea *Geophys. Res. Lett.* **29** 76-1–76-4

Weijer W *et al* 2019 Stability of the Atlantic meridional overturning circulation: a review and synthesis *J. Geophys. Res. Oceans* **124** 5336–75

Melody Sandells and Daniela Flocco

Chapter 8

Thoughts for the future

The study of the cryosphere is a useful and necessary tool to help us in the understanding of greater issues related to the Earth's climate and its impact on the human population. This falls into a greater framework of studies, assessments and mitigation strategies that most countries agreed to implement to face the challenges imposed by our rapidly warming climate.

The 26th UN Climate Change Conference of the Parties (COP26), which took place in Glasgow, in 2021, 'reaffirms the Paris Agreement temperature goal of holding the increase in the global average temperature to well below 2 °C above pre-industrial levels and pursuing efforts to limit the temperature increase to 1.5 °C above pre-industrial levels'. It also 'recognizes that the impacts of climate change will be much lower at the temperature increase of 1.5 °C compared to 2 °C and resolves to pursue efforts to limit the temperature increase to 1.5 °C, recognizing that this requires meaningful and effective action by all Parties in this critical decade on the basis of the best available scientific knowledge, reflecting common but differentiated responsibilities and respective capabilities in the light of different national circumstances'.

COP26 used the outcome of the Intergovernmental Panel on Climate Change (IPCC) reports to agree on common guidelines on mitigation strategies to follow. The IPCC report on the polar regions (Meredith *et al* 2019), is a detailed document, to which we refer for further insights. A summary examining the impact of a temperature increase of 1.5 °C on the Earth's natural and human systems can be found on the NASA website.

Here we briefly focus on the use of models together with extensive use of observations gathered *in situ* and from satellites. Models are one of the main tools that the IPCC uses to give information and directions. Depending on the process studied, models have different degrees of precision. When talking about global circulation models, used for the study of global-scale processes, compromises must be accepted, often in term of spatial resolution. When looking at

doi:10.1088/978-0-7503-3647-5ch8

Arctic sea ice, for example, predictions have an uncertainty due to the intrinsic characteristics of the model used—the level of detail applied. Knowledge of the power and limitation of model prediction is key to the interpretation of the results produced.

A 2016 study performed a number of experiments to assess the level of predictability of the timing of an ice-free summer Arctic. They concluded that, due to internal variability alone, the *uncertainty* of their prediction could not be less than *21 years* (Jahn *et al* 2016). This kind of study, which rigorously assesses the accuracy of the results, is a perfect example of the good practice that will form the basis of further advances in our understanding of the Earth's systems.

Schwierz *et al* (2006) divides the possible model biases into two categories.

1. Predictability of the climate system—a number of causes intrinsic to the system studied can impact its predictability:
   - Climate systems are characterised by natural variability due to the nonlinear interaction between their components.
   - Imperfect initial conditions.
   - Imperfect boundary conditions.
2. Model error (imperfect models):
   - Limited representation of physical processes.
   - Limited knowledge of processes.
   - Structural uncertainty.

The efforts of the scientific community are aimed at understanding and assessment of biases belonging to the first category of error sources, while the continuous improvement of climate models aims to reduce uncertainties in the results due to technical limitations. But, we need more people to work on this. The rapid climate changes we are experiencing add an extra layer of difficulty to the ones listed above and means this is a critically urgent problem to tackle. For further reading on the challenges in cryospheric sciences, we suggest a 2017 study (Hock *et al* 2017), where the authors outline several outstanding issues that the science community is addressing and will continue to do so in the future with the aid of new data-collection techniques and improved models. New techniques based on machine learning are also coming to the aid when working towards understanding of complex systems and large amoun of data such as satellite data (Lin 2021). This is a hot topic for research, but is not covered here.

There is still so much more to be done to understand how our cryosphere has changed in the past and will change in the future; to quantify, understand and reduce errors in both our models and observations; to draw parallels between similar processes and to use interdisciplinary and collaborative approaches to link how the biosphere responds to the physical and chemical changes occurring around us; to recognize how societal and political behaviour can influence the trajectory of our future. This primer is designed to give a grounding in the physics of the cryosphere as a stepping stone to your future research. We cannot wait to learn about what you discover!

References

Hock R, Hutchings J K and Lehning M 2017 Grand challenges in cryospheric sciences: toward better predictability of glaciers, snow and sea ice *Front. Earth Sci.* **5** 64

Jahn A *et al* 2016 How predictable is the timing of a summer ice-free Arctic? *Geophys. Res. Lett.* **43** 9113–20

Lin L 2021 A review of deep learning for cryospheric studies *Deep Learning for the Earth Sciences* (New York: Wiley) ch 17 pp 258–68

Meredith M *et al* 2019 Polar regions *IPCC Special Report on the Ocean and Cryosphere in a Changing Climate* (Cambridge: Cambridge University Press) ch 3

Schwierz C *et al* 2006 Challenges posed by and approaches to the study of seasonal-to-decadal climate variability *Clim. Change* **79** 31–63

www.ingramcontent.com/pod-product-compliance
Ingram Content Group UK Ltd.
Pitfield, Milton Keynes, MK11 3LW, UK
UKHW051849100926
8556IPUK00005B/546

9 780750 336451